GO TO MARKET LIKE YOU MEAN IT

GO TO MARKET LIKE YOU MEAN IT

THE TACTICAL FIELD GUIDE FOR SCALING SAAS REVENUE

JOHN BRAZE   ASH ARCHIBALD   CARLOS VADILLO

Mir

PRAISE FOR GO TO MARKET LIKE YOU MEAN IT

"*Go to Market Like You Mean It* opens a window into a world I don't usually inhabit. What struck me is how deeply its principles reflect what leadership is all about: clarity and discipline about what matters, care for people and relationships, and creativity in how you get things done. It's a practical field guide for anyone trying to turn good ideas into real-world results."

STEW FRIEDMAN

Wharton Professor and Author of 'Total Leadership'

———

"This book effectively connects strategy and execution. It's not pedantic. It's a manual and diagram for action for every founder who cares about business outcomes. *Go to Market Like You Mean It* is a compelling and rare mix of intellectual honesty, operational depth, and empathy for the founder's grind. Every chapter turns experience into leverage."

MARK DE SANTIS

CEO Bloomfield Robotics, Co-Founder & CEO Roadbotics, and Carnegie Mellon Adjunct Professor

———

"Working across media, sports, and technology has shown me that execution wins. The teams that succeed are the ones that align vision, people, purpose, and performance. *Go to Market Like You Mean It* brings that same discipline to startups. It gives founders a path forward to converting creativity into structure and momentum into measurable growth. Every emerging company can benefit from the clarity this book provides, and ambitious thinkers looking to leverage strategy and growth will leave smarter after reading this."

SANDY SHARP

Media Executive and Technology Advisor

CONTENTS

APPENDICES

COMPANION RESOURCES

Use the QR Code to access the companion resources for Go to Market Like You Mean It.

mirmeridian.com/resources

Download templates, worksheets, checklists, GTM-focused AI bots, and planning tools referenced in the book to help you apply the content directly to your go-to-market. The materials are updated periodically. Register for updates on our website's Resources page.

IS THIS BOOK FOR YOU?

Quick Self-Test

- You cannot explain why **some deals close and some stall.**
- Your CRM shows activity, yet **replies and meetings are low.**
- **Time to first touch is slow.** Leads leak before Sales engages.
- **Your ideal customer profile is a slide in a presentation,** not a list of real accounts that convert fast at full price.
- **Messaging only works when the founder speaks.** It fails when others use it.
- **Outbound is busy, but ineffective.** Reply rates are under 2%.
- **Pricing is a guess.** Discounting is the default to get deals over the line.

- **Handoffs are messy.** Marketing blames Sales. Sales blames Product.
- **Sales forecasts swing week to week.** You do not trust the numbers.
- You have fancy **dashboards that no one uses to make decisions.**
- **Expansion is accidental.** Customer Success chases tickets, not outcomes.
- **You keep buying tools** to fix process problems.

If you answered yes to one or more of these questions, and you want to fix it, we wrote this book for you.

What This Book Does

This book helps you install a complete RevOps backbone in 30 days: defining your ideal customer profile, sharpening your messaging, aligning channels, tightening processes, setting pricing, and creating the weekly operating rhythm that keeps it all real. It makes performance unmistakable with a numbers-first approach. When a motion falters, you know exactly how to fix it; when a motion works, you can double down confidently. Every chapter includes practical tools and checklists you can use immediately and field-tested steps you can run weekly, not guesses or abstract stories.

Who This Book is For

This book is written for founders, CEOs, operators, and sales leaders in B2B SaaS who value "steak, not sizzle," and want a system their team can run without them. It's for teams that need clarity on pipeline quality, sales cycle, win rate, and expansion. And it's for leaders ready to align Product, Marketing, Sales, and Customer Success around one unified operating plan.

Prerequisites

You'll get the most value from this book if you have evidence of product-market fit—real usage, renewals, or referrals. You should have a handful of paying customers and a solid working idea of your ideal customer profile. Most importantly, you need enough motion to generate data, even if that data is messy.

Who This Book is Not For

This book is not for teams looking for a content marketing course, a sales training manual, or a brand design guide. It's also not for founders or CEOs who refuse to use, update, or enforce a CRM and believe a spreadsheet and a stack of sticky notes are enough to track leads.

———

Start here if you are ready to replace hustle with a data-driven, repeatable go-to-market system. The first wins will arrive quickly.

VOICES FROM THE FIELD

THE AUTHORS

Before we get into frameworks, we want you to hear from people who've lived this journey. The voices ahead come from founders and operators at every stage of scale: some who've crossed the $10M mark, others still in the trenches, and seasoned advisors who've helped dozens do the same. They've built and refined the systems we'll unpack in this book, sometimes through painful trial and error. Each understands the difference between revenue that happens to you and revenue you can predict, scale, and defend. Their insights set the stage for what follows.

———

AMANTHA BAGDON
Founder & CEO Rxpost

I never set out to start a tech company. I just wanted to solve a problem I saw every day behind the pharmacy counter. Piles of perfectly good medication are going to waste on pharmacy shelves, while at the same time, patients are struggling to find the medicines they need. The waste wasn't from a lack of caring; it was from a lack of connectivity, data being siloed, or, more often, information being stored on paper instead of electronically.

Building RxPost means learning to think like an operator and an advocate simultaneously. We aren't chasing startup clout; we are fixing real workflows. Every choice about compliance, inventory, or logistics has real consequences for people. *Go to Market Like You Mean It* hits home because it speaks the language of people with lived experience and profound founder-market-fit, balancing empathy with execution.

The book doesn't glamorize growth. It focuses on alignment and how to turn purpose into process and processes into measured outcomes. For us, that means turning compassion into code and building trust one transaction at a time.

Founders often chase disruption, but most industries don't need more disruption. They need better connection. This book gets that. It reminds us that the real leadership work happens between the strategy slide and the day-to-day work, where people actually feel the impact of what we build.

This is more than a go-to-market book; it's for anyone trying to build something that connects people and creates lasting value. It's a reminder of why we build in the first place: with clarity, integrity, and care.

———

MARK DE SANTIS
CEO Bloomfield Robotics, Co-Founder & CEO Roadbotics, and Carnegie Mellon Adjunct Professor

I enjoy the challenge of making something from nothing with nothing. To me, that's the definition of entrepreneurship. However, things can go very wrong sometimes in hard-to-predict ways. In my experience, one big reason is that I did not think through and plan how, when, and where to go to market.

The first time I read *Go to Market Like You Mean It*, I felt like I was sitting with operators who've actually been through it. The authors are pragmatic and thoughtful, with deep experience in sequencing, data integrity, and truth-telling inside the business; the unglamorous stuff that separates disciplined companies from hopeful ones.

At Bloomfield Robotics and earlier startups, I learned that execution always beats enthusiasm. You can have the best vision in the world, but if you don't have a clean system for turning that vision into a pipeline, revenue, and feedback, you're just guessing. This book outlines how to align leadership, data, and process so your go-to-market motion compounds over time instead of collapsing under its own weight.

It's rare to find a playbook that's both intellectually honest and battle-tested. *Go to Market Like You Mean It* is that book. It's not just advice; it's a reminder that the real leadership work happens in the trenches, not in the pitch deck.

———

CORY BRAY

Co-Founder Clozeloop, Sales Enablement Expert, and Author of Eight Sales Books

I've spent my entire 20-year career working in and around B2B sales teams, and I can confidently say that Revenue Operations in the early days is almost always a disaster. But it doesn't have to be...

Here's what typically happens: executive teams exhaust themselves building a great product, then build a team to sell and support it. But when it comes to operations? They wire some stuff up with duct tape and WD-40 and call it a day.

Anyone *can* set up a CRM on a Sunday while watching football.

Anyone *can* set up a reporting package in a spreadsheet.

Anyone *can* buy and implement other tools to support the business.

This "good enough" mindset is a trap. What feels functional today *will* become years of random problems and re-work that surface at the worst possible times, usually when you're trying to scale or close a big deal.

This book solves that exact problem. It gives early-stage leaders a clear roadmap for doing things right from day one, without the mess that needs unwinding later.

Better yet? It doesn't take hundreds of hours or require hiring additional W-2 employees to get it done.

————

KRISTEN NUNERY
Founder & CEO Illumend

Building illumend has meant solving the same challenge our customers face daily: turning complexity into clear execution. As we scaled, vision was never the problem — alignment was. As the team expanded, the stakes rose, and the space between strategy and action stretched thin; our real test wasn't ambition. It was precision.

I've led illumend through early traction, product pivots, and that messy middle where revenue is real but systems are still catching up. At that point, you don't need more ideas. You need discipline with a framework that holds steady under pressure and keeps everyone rowing in rhythm when growth gets loud.

Go to Market Like You Mean It is written for leaders who've lived that. The book doesn't waste time with hypotheticals. It dives straight into the decisions that make or break momentum: which segments to prioritize, who owns what, how to tune systems, and how to make your go-to-market engine repeatable. It's not for startups sketching on napkins, it's for CEOs already in motion, determined to scale without spinning out.

The power of this book is in its practicality. Each chapter is a working playbook, something you can apply that week and see real results. For any B2B CEO steering through growth, this isn't another abstract strategy guide. It's a sharp, honest manual built to raise your game and tighten your execution — one clear decision at a time.

TIM BRADY
CEO Colligo

When I acquired Colligo in 2019, the world had not yet fully accepted that business records belonged in the cloud. The vision was clear: make enterprise-grade email governance and compliance simple, compliant, and secure. The challenge was alignment. How do you evolve a 20-year-old on-premise company into a cloud-first platform without losing what made it trusted in the first place?

For us, reinvention was not optional. It meant rebuilding Colligo from the ground up, and ensuring compliance and efficiency gains were not a burden but a built-in outcome for customers. When the pandemic accelerated digital transformation, we were already there. But scaling was never just about chasing growth. It was about precision and readiness. Those are the same disciplines this book emphasizes.

What resonates most for me in *Go to Market Like You Mean It* is its insistence on clarity and structure. Ensuring everybody is rowing in the same direction and the structures are in place to provide a foundation for scaling is critical.

This book reflects that same mindset: disciplined execution and data-driven insight. Founders often confuse activity with execution. They believe speed is everything, yet speed without structure multiplies mistakes. Doing things the right way does not slow you down; it prevents the chaos that comes from rushing. Real progress is quiet, methodical, and measurable. This book drives that point home with clarity and experience.

———

THE AUTHORS

The stories you've just read remind us that every founder's path looks different, but the discipline behind success is universal. Alignment, clarity, and structure turn intent into impact. These voices prove that growth isn't luck; it's leadership in motion. As you move forward, take their lessons as proof that what's hard is often what works.

Up next: how to use this book to build your own version of that discipline, step by step.

HOW TO USE THIS BOOK

"You do not rise to the level of your goals. You fall to the level of your systems."
— *James Clear*

Insight

This tactical field guide provides tried and tested processes for building and running Revenue Operations (RevOps) in a B2B SaaS company. Your data, the length of your tests, sales cycle, and prices depend on your specific offerings, customers, and deal sizes. The goal is to master the concepts and install a system that fits your business.

Sprints, Stages, Handoffs, Cycles, and Other Jargon We Use Too Much

We use a lot of jargon, even though we've heard that's not the best practice when writing a book. We did it anyway. You and your team will do it, too. The important thing is that your entire team is crystal clear on what the terms in your specific GTM dialect mean.

There is an extensive glossary at the end of the book, in case you encounter unfamiliar terms (or many of them). Make your own glossary, and insist that your team aligns on the terms.

Put This Book to Work

- **Start with Section 1** to frame why GTM and RevOps exist and how to measure progress.
- **Move through Sections 2 and 3** to assemble the engine: ideal customer profile (ICP), messaging, channels, pricing, process, RevOps, and hiring.
- **Use Section 4** to operate: reporting, leadership cadence, and AI-readiness.

Work It Weekly

- Pick your most significant constraint: pipeline quality, win rate, sales cycle, or expansion.
- Apply one tool this week. Measure one metric. Decide on one change.
- Review results next week. Keep what works. Cut what does not. Repeat.

Use The Fewest Metrics Required

- Some teams need five metrics. Others need 50. Track only what you use to make decisions (or think you might need soon).
- Don't know where to start?
- Start with four metrics: **reply rate, meeting rate, win rate,** and **sales cycle.** As you mature your processes, add **price realization** and **revenue retention** (NRR and GRR).
- Build a simple triage & hygiene dashboard in your CRM or spreadsheet (keep it simple, automate it). Start with pipeline deals with:
  - No activity in the past 14 days, or
  - No actionable next steps recorded, or
  - An expected close date in the past, or
  - A deal value greater than the median.
- Review those deals honestly. If you aren't working them, have no next steps, and keep moving the close date, then accept reality and move the deal to Closed-Lost. Don't clutter your pipeline with fantasy. You cannot sell product to fantasy customers.

Look at metrics and deals every week. Hold your team accountable. Perform triage & hygiene. Measure the four metrics above; graduate to the canon as your stage gates hold up. Don't hallucinate. Fantasy is fatal.

Pipeline Stages and Exit Criteria

- Each pipeline stage needs a short checklist to proceed. For example, for "Qualified," the problem is confirmed, the champion is identified, and the next step is timeboxed.
- Add the **proof of value** you expect to show to the lead (e.g., ROI math, reference call, infosec compliance, pilot plan).

Understand and Develop Your Sales Models

- Product-led growth (PLG) motions need clear value events, product-qualified lead (PQL) or product-qualified account (PQA) rules, and fast handoffs.
- Inbound needs fast SLAs, compelling content, and effective capture points.
- Outbound needs tight lists, triggers, and message tests.
- Enterprise motions need fewer, deeper touches,

stricter exit criteria, and involve more buyer-side stakeholders.

Run Experiments in Short Timeboxes

- If you have a short sales cycle, many tests can be completed in two weeks. Longer sales cycles may require longer timeboxes, but running shorter experiments is better whenever possible.
- Define a pass-fail threshold before you start.
- Change one variable at a time: design before scale, message before outbound, process before hiring, data before dashboards.

The Two-Week Rule (and when to bend it)

Two weeks forces focus. In two weeks, you can:
- Test a subject line and measure open/reply rates.
- Launch a landing page and see if anyone clicks "Book a Demo."
- Run 100 outbound calls and track conversations.

But what if your sales cycle is 180 days?
- Don't wait six months to learn. Shift your metric, not your cadence.
- SMB SaaS (short cycle): measure sign-ups or trials in two weeks.
- Midmarket (medium cycle): track demos booked or MAP completion every two weeks.
- Enterprise (long cycle): log exec meetings, proofs of concept started, or decision-maker engagement.

The principle: Every two weeks, you should have new facts. The metric may change with the sales cycle, but the rhythm never does.

Develop a Weekly Operating Rhythm

For example,

- **Monday:** agree on goals, experiments, and owners.
- **Daily:** work your GTM playbook and monitor that SLAs are met successfully.
- **Friday:** review the numbers, tie up loose ends, and review lessons learned.

Our Experience

We have seen teams scale with fewer than 10 metrics and a simple cadence. We have also seen dashboards with 60 charts that are never used. The winning teams treat RevOps as the backbone that keeps their plans honest. They make small changes fast and measure what those changes do to the metrics.

Putting It Together

RevOps is not prescriptive. It is a discipline you build. Your GTM strategy is idiosyncratic. Use this book to create a RevOps backbone that fits your customers and model, then toe the line. Be disciplined about capturing the data you deem crucial. Everyone participates, from the CEO to the sales development rep (SDR). The result is a system that sells without needing the CEO or the founder in every meeting.

Companion Resources

Start with the Quick Start Worksheets (WS01–WS05). Use our AI bots to **validate your SOM**, assess your **GTM readiness,** or for a general **business evaluation.** Use them to stand up your RevOps backbone in the next 30 days.

Download templates, worksheets, checklists, GTM-focused AI bots, and planning tools referenced in the book to help you apply the content directly to your go-to-market. The

materials are updated periodically. Register for updates on our website's Resources page.

The Resources are available by using the QR code in the Companion Resources section at the beginning of the book.

INTRODUCTION

"Startups don't starve, they drown."
— *Chris Dixon*

From our experience, most growing companies don't fail because they lack ideas. They fail because they chase too many things at once and never build a system that turns effort into revenue. Tech founders are savvy, have a product vision, and the market has potential, but none of that matters without a go-to-market strategy that works under pressure. This book is for founders who don't want to drown.

This is a field manual for leaders ready to turn products into pipelines and pipelines into durable businesses. Many teams work hard for a long time without success because they lack a sales system to attract, inform, and convert customers. Instead of designing GTM with intention, they improvise, and the cracks show up in lost revenue, morale, and investor trust. This book guides you through building that system using straightforward tactics proven in the field by people who've built and sold successfully. You'll learn the fundamentals of revenue operations, including vital definitions, metrics, service-level agreements, and scorecards to track progress and improve performance.

Your company might be pre-seed and closing its first 10 deals, or at $10 million ARR and wondering why nothing scales beyond your own effort. You might even be at $25 million, about to hire a GTM team, and realizing the product story only works when you're in the room. This book is for CEOs, founders, and sales leaders who need a system that works without burning out their people or their cash. It's also for investors who evaluate GTM maturity beyond revenue alone.

The chapters follow the real work of building a motion that sells:

- Defining the ICP
- Getting replies
- Running a sales motion
- Building a process before hiring
- Tracking the proper metrics with a weekly cadence
- Creating a pricing strategy that sells
- Hiring a team that doesn't fail
- Leading after you find traction

Each chapter ends with frameworks and tools you can apply immediately. Startups rarely die from lack of opportunity—they die from noise, long sales cycles, and unclear revenue paths. Methodical go-to-market isn't a department or a hire;

it's a mindset. It's how you survive. Approach it like an adventure: assess your skills, choose your methods, and know when to push forward or make adjustments.

If you started a company hoping sales would "just happen," you're not building a company; you're playing the lottery. This book shows you how to stop guessing and build the system that gives your product a real shot at winning. We wrote this as a conversation with you, not a lecture. You're doing something hard, and we plan to be with you on the journey.

meet.mirmeridian.com/gtm-book-meeting-qr

If something comes up that you want to discuss, scan the QR code below to schedule a conversation with us about this book or your own go-to-market challenges.

SECTION I

THE FOUNDATION OF GTM

1

GTM IS A SURVIVAL SKILL

"Vision without execution is hallucination."
— *Thomas Edison*

Examples and Metrics

Throughout the book, we often give examples like "do this for two weeks," or we say "hit this metric." Remember that these are examples and not every business will set the same metrics to monitor, KPIs to hit, or have similar growth rates. Every company has different products at vastly different price points and sales cycles. If you'd like to talk through your particular metrics, feel free to set up a meeting with us using the QR code in the Introduction. We'd be happy to be a sounding board.

Insight

Most start-ups do not fail for lack of effort. They fail because they confuse activity with a system. GTM is not a hire, a pitch deck, or a quarterly campaign. It is the operating system that turns strangers into customers and customers into advocates in a way that works when you are not in the room. If revenue depends on a founder's cameo or an account exec's mood, you do not have a GTM strategy. You have a house of cards.

GTM in Five Moves

1. GTM is the system that turns strangers into customers.
2. It defines who you target, what you say, and how you reach them.
3. It specifies what happens when they respond, so the path to a win is teachable.
4. It measures why you win or lose and feeds that into pricing, product, and messaging.
5. If it does not work without you, it is not GTM. It is your hustle, and it is not scalable.

Practical Walkthrough

Build a simple, testable GTM system in five moves.

1. Define who and why

Identify your ICP, the segment you can actually reach now. List three ICP buying triggers you can detect.

2. Define buyer-language messages

Identify one pain, one promise, and one proof of value that you can demonstrate. Write three message subject lines and A/B test them. Keep email messages under 125 words and social media direct messages even shorter. Measure reply rate, not open rates. Open rates reported for email are often much higher than reality due to bot activity. Don't get excited about open-rates. Replies are what matter.

3. Choose channels and SLAs

Pick one outbound channel, PLG, or inbound capture point (as applicable) and own them. Define SLAs for your response time to user activity. For example, first touch in five minutes for inbound, PQL to a human in less than two hours for PLG, or follow-up in less than 24 hours for outbound replies.

4. Map the path to a win

Document each stage of the deal: qualify, prove, align, and commit, and make the process visible to both sides. Share a Mutual Action Plan (MAP) with specific dates and named owners for your team and the customer. If the buyer has not accepted a MAP within 48 hours of discovery, pause the deal until they commit to a defined plan that leads to a closed-won deal.

5. Execute and iterate

Track four core weekly metrics: reply rate, meeting rate, win rate, and sales cycle. Adjust only one activity at a time, such as an email style or marketing message, and roll it back if it doesn't improve the target metric within two weeks. If your sales cycle exceeds 60 days, maintain the two-week experiment but evaluate leading indicators instead, such as qualified positive replies, scheduled executive meetings, MAP adoption, or proof of value started.

What to Avoid

Don't run outbound at volume until you confirm which messages generate replies. Avoid hiring account executives (AEs) before you can teach a repeatable sales path. Eliminate dashboards that do not drive specific decisions.

Example

We worked with a seed-stage company with an ARR of about $800K that ran only on inbound. CRM leaks let MQLs die before the first touch. We fixed their RevOps to capture and segment every lead, set SLAs, and added a light outbound motion only after the message produced replies. The time to first touch dropped to minutes, the meeting rate climbed, and the win rate improved. They tripled ARR in 12 months.

"You know the difference between hitting .250 and hitting .300? It's 25 hits. 25 hits in 500 at-bats is 50 points. That's about one extra hit a week. It's the difference between a .250 hitter and a .300 hitter. It's one extra hit a week and then you're in Yankee Stadium."

— *CRASH DAVIS (BULL DURHAM)*

The difference between mediocrity and greatness isn't massive. Sometimes it's just one more win, one more closed deal, one more meeting booked, an email replied to, or a customer converted per week.

The power of RevOps is its ability to operationalize your GTM strategy. In the early days, dedicating just a few hours per week to RevOps can be the difference between hitting $10M ARR or languishing in the high six figures forever. For those who break $10M, RevOps gets you to nine figures as you move from growth to scale. **RevOps is one of the highest ROI activities your company can invest in.**

Our Experience

Carlos saw firsthand how even real traction can slip away without a system. His start-up, Agribots, had promising pilots and sincere buyers interested. But without a clearly defined ICP, mapped conversion flow, or visibility into deal signals, the team could not replicate success or pinpoint breakdowns. Carlos poured energy into every motion, but the absence of structure cost the company time. That experience drives his conviction today: you need a system early, and you need to measure what matters from day one.

We've seen the inverse, too. We advised a product management start-up full of hustle, attending conferences, running events, and flying nationwide. But they hadn't validated market size, segmented their outreach, or agreed on a real ideal customer. Their pricing and positioning demanded a tightly qualified market, but the numbers didn't support the model. They weren't lacking effort; they were missing alignment. That company nearly ran out of runway, not for lack of hustle, but for lack of strategy.

Putting It All Together

Your job as the founder, CEO, CRO, or VP of Sales is not to prove that selling is possible. Your job is to make selling teachable. Get the order right: design before scale, message before outbound, process before hiring, data before dashboards. The rest of this book follows that order, so you know what to do now and what to ignore until later.

Companion Resources

Worksheets WS01, WS02, WS03, WS04, and WS05 (the "Quick Start Pack") support Chapters 1 and 2.

Download templates, worksheets, checklists, GTM-focused AI bots, and planning tools referenced in the book to help you apply the content directly to your go-to-market. The materials are updated periodically. Register for updates on our website's Resources page.

The Resources are available by using the QR code in the Companion Resources section at the beginning of the book.

2

MEASURING PROGRESS
THE RIGHT WAY

*"Numbers have an important story to tell. They rely on you to
give them a clear and convincing voice."*
— *Stephen Few*

Insight

Revenue can hide fragility. A few big deals, a friendly
network, or founder heroics make results look better than
the system actually is. Pipeline stages remove the guesswork
when they tell you what to prove now, what to ignore until
later, and when to change headcount or spend. The
company will scale noise if you do not define the exit criteria
for each pipeline stage.

Practical Walkthrough

Use pipeline stages and exit criteria to make decisions and

assess progress. Set your GTM targets based on your GTM Maturity Stage.

Define your company's current GTM maturity stage. It helps with expectations, decision-making, and prioritization. It's not defined by revenue. We've worked with companies with eight-figure ARR that still depend on the founder to close deals.

Further, set your criteria for entering the next GTM maturity stage. This is long-term planning. Define the thresholds that prove you are ready for the next level. Keep them simple, numerical, and observable.

GTM Maturity Stages

ZERO TO SOMETHING

PROVE: Real buyers will pay, and at least one repeatable path to a win exists.

Example metrics:

- Closed won 3–5 cold ICP accounts.
- Median discount ≤10%.
- Meeting → Pipeline ≥20%.
- Pipeline → Win ≥20%.
- Median sales cycle is stable for 4 weeks.
- One proof of value (e.g., demo, pilot, testimonial) that buyers accept.
- Value delivered in 30 days.

SELLING BEFORE SCALING

Prove: The motion repeats every time, and handoffs hold consistently.

- ≥70% of wins following documented steps for two straight months.
- MAP used on ≥80% of late-stage deals.
- SLA hit rates ≥90% (form → first touch, lead → meeting).
- Stage accuracy ≥95%. Median discount ≤10–12%.
- Reply and meeting rates are stable without founder assist.

HIRING WITHOUT BREAKING

Prove: AEs (non-founders) can win deals repeatably.

- Two new AEs at 60–90 days each hit ≥80% of baseline win rate and ≥80% of SLA compliance.
- Founder-involved deals ≤20% of wins.
- Forecast accuracy within ±15%.
- Enablement playbook in use.
- No lift in median discount or sales cycle length.

SCALING WITH CONFIDENCE

Prove: Efficiency holds at volume.

- No single channel >60% of meetings for two quarters.

- CAC payback ≤12–18 months (stage-appropriate) on trailing three months.
- ±10%. forecast accuracy.
- ≤10% median discount
- The capacity model hit rate is within ±10%.
- Churn and NRR are stable by cohort (if applicable).

Install a RevOps Backbone

Build a RevOps backbone as the minimum set of metrics, CRM fields, SLAs, and scorecards that connect your funnel from end to end: stable at the core and flexible at the edges. Capture only essential data such as accounts, contacts, activities, opportunities, and products. Define lifecycle SLAs for every handoff, including form to first touch, lead to meeting, meeting to opportunity, and proposal to close. Maintain hygiene by requiring a next-step date on every active deal, accurate stage updates, and standardized lost-deal reasons — always document why you lost deals. Be specific. "Loss to a competitor" is not a reason. If you lost a deal, there was a reason: Price? Product? Process?

Pipeline Review and Score Goals

- GREEN: if it's met for two consecutive weeks.
- YELLOW: if trending.
- RED: if missed.

Name a bottleneck to fix and an experiment to implement, then assign owners.

Stage Gate Control Decisions

- **Hiring:** do not add AEs until the "Selling Before Scaling" targets are held for a reasonable period.
- **Channels:** add AEs only after the current primary channel hits its target repeatedly.
- **Pricing:** run experiments, then lock a package for the next cycle to confirm validity.

Keep-Kill Rules

- Keep a change only if it positively moves the target metric within your defined timebox.
- Kill a channel or segment if reply quality or meeting rate declines for two consecutive sales cycles, or as soon as you know.
- Roll back a pricing test if the discount rate exceeds your allowable limit.

Example

A Series A team was chasing five segments with one product and closing almost nothing. Sales cycles stretched past 90 days, and win rates stayed under 10%.

They pulled back on outbound for two weeks and worked only inbound and warm leads. During that window, they reviewed every closed won deal and found a single repeatable use case: midmarket finance teams automating compliance.

They rebuilt their outbound motion around that use case, enforced "keep-kill" rules, and required MAPs and stage-based advancement. Within six weeks, the win rate doubled, the sales cycle dropped by 20%, and reps started hitting quotas.

Our Experience

We have been called into companies that look strong on paper but crumble under inspection. In one case, the founder believed they were ready to scale on the strength of a packed pipeline and enthusiastic demos. The data said otherwise. Deals died. Nobody owned the economic buyer, and there were no MAPs. It was marketing-led volume without ICP discipline. We helped the company develop deal stage criteria, put SLAs in place, and enforced every late-stage deal to have a MAP within 48 hours. The result was fewer steps, cleaner handoffs, and an improved win rate.

Putting It All Together

GTM maturity stages break ambition into actionable steps. Chapter 3 will size the obtainable market you can sell into (SOM). Chapter 4 will sharpen the ICP that closes at speed. Chapter 5 will improve replies. Chapter 9 will keep the data honest. Use the GTM maturity stage to decide what you need to accomplish next. If a change breaks a RevOps goal, fix the motion. If you do something and the result is not what you expected, undo what you did. Pilots call it airmanship. Don't wait and hope the plane won't crash. Fly the plane.

"Don't let yourself get so far behind the airplane that you don't know what's going on or what to do next."

— CHESLEY B. SULLENBERGER

Companion Resources

Worksheets WS02, WS29, WS30, and WS36 support Chapter 2. Use them to establish baseline metrics, set stage gates, and perform weekly reviews.

Download templates, worksheets, checklists, GTM-focused AI bots, and planning tools referenced in the book to help you apply the content directly to your go-to-market. The materials are updated periodically. Register for updates on our website's Resources page.

The Resources are available by using the QR code in the Companion Resources section at the beginning of the book.

3

MARKET SIZING THAT DOESN'T LIE

"All models are wrong, but some are useful."
— *George Box*

Insight

A slide showing an enormous TAM does not create revenue. You sell to people you can reach and win this year and next. That is your SOM, which must be time-constrained (the next four years is a reasonable anchor). If you build GTM on top of wishful TAM math, you will hire the wrong team, pick the wrong channels, and run out of runway. Size the market from the bottom up, accounting for behavior and capacity, so your plan reflects reality, not optimism.

We call this "turning on reality mode." TAM calculations do not presume you can ever reach them (which makes sense because you won't and can't sell up to your TAM). Top-down TAM calculations are like assuming you already have the universe in your hand and need to figure out how to hold as much of it as possible. The real world is not involved in that fantasy.

Practical Walkthrough

Build a bottom-up SOM and validate demand and your capability to meet it before you scale. Unlike TAM, your SOM assumes you have nothing, and you need to figure out how to get scrappy and catch pieces of the universe with your existing real-world resources.

For example, assume it's a fact that you can close a deal with every ICP prospect you meet. Now ask:

- How many prospects can you find and convince to take a meeting?
- Will the prospect show up for the meeting?
- How many prospects can you meet in a year?
- Does the prospect have a budget?
- Is your sales cycle one week or one year?

That's what reality mode turned on looks like. It's the difference between TAM and SOM.

Step 1: pick one ICP segment to model.

- Start with prospects who close fast and pay full price.
- Write down three observable triggers you can detect that signal they are ready to buy.

Step 2: build a bottom-up SOM (in reality mode).

If you don't have enough data about these metrics, make assumptions, then start trending real data and improve it. It's never too early to collect information about your sales process and customer behavior. One method of calculating SOM is to estimate as accurately as possible how many ICP accounts are available, and then calculate your capacity to sell to them. There are many valid methods to calculate SOM. The important concept is that SOM is constrained by time and resources. Resources include cash, people, and the ability to design and ship product, among other constraints. **These constraints are real, and they impact how fast you can scale.**

Variables:

- Accounts in segment (A): total ICP accounts that meet your triggers.
- Reachable per month (R): how many you can touch with quality.
- Meeting rate (M): meetings ÷ touches.
- Conversion rate (C): opportunities ÷ meetings.
- Win rate (W): closed won ÷ opportunities.
- Average selling price (S): asking price minus typical discount.

Monthly ARR capacity = (R × M × C × W × ASP)

Annual capacity = (monthly capacity × 12)

Compare to AE capacity: AE capacity = (meetings per AE per week) × (opportunity yield) × S

EXAMPLE

10,000 total accounts

1,000 reachable per month

10% meeting rate

30% pipeline rate

25% win rate

22,000 average selling price

Capacity = 1000 × 10% × 30% × 25% × $22K = $165K new ARR

Out of 1,000 quality touches, you'd expect 100 meetings, 30 opportunities, and ≈7 or 8 closed deals worth about $22K each, totaling $165K in new ARR.

Choose the Right Sales Motion

1. Choose a sales model that matches your current stage.

In the early phase, a founder-led model works best when the market is narrow, deals are complex, and the goal is learning. As reach expands to roughly 1,000–10,000 accounts, an SDR plus AE model becomes viable once the path is teachable. Product-led growth (PLG) suits high-volume, low-friction motions with strong activation to paid users, adding sales assist for large opportunities. A channel or partner model should only be used once you can reliably win direct deals in the same segment in which your channel partners operate.

For new roadmap features, test demand before investing in development. Run a single, timeboxed experiment for one segment with clear pass and fail thresholds to validate interest before you build.

For example:

- **Fake door:** offer a feature or package on a landing page. Implement it if the click-through to "get access" is ≥2–5% of qualified visitors; otherwise, kill it. Be clear to the buyer what exists and what doesn't. Don't sell vaporware.
- **Painted-door pricing:** show two tiers. Implement if ≥30% select the higher-value tier.
- **Concierge trial:** manually deliver the outcome for 5–10 targets. Implement if ≥3 pay or sign letters of intent with clear value metrics.
- **Smoke ads:** targeted ads to a landing page. Implement if the qualified-lead cost fits your CAC payback target.
- **Pre-order or deposit** (where appropriate): implement if at least 5–10% of late-stage evaluators commit. Ethics: ensure clear refund terms.

2. Keep or kill rules

- Keep a segment or offer only if reply, meeting, and

win rates meet your targets for two consecutive experiment cycles.

- Kill a channel if cost drifts up and reply quality drops for two experiment cycles.
- Roll back a price test if the discount rate rises or the sales cycle grows.

Example A: TAM Fiction

Claimed market: "every firm with 10+ accountants."

Actual reachable market: A few hundred firms with budget, urgency, and a buyer.

The bottom-up SOM showed that the addressable revenue could not support the price and heavy CAC motion. The company stopped events and travel, re-aimed at a narrower segment, and stopped significant cash burn.

Example B: HR Analytics Reality

Claimed market: 4,000 ICP accounts in midmarket HR.

Actual reachable market: 500 accounts.

Meeting rate 10%. Pipeline rate 25%. Win rate 30%. ASP $12K.

ARR capacity = (500 × 0.10 × 0.25 × 0.30 × $12K) = $45K per month

Annual capacity is ~$540K from one segment and channel. That is a plan, not a fantasy.

Our Experience

We have rebuilt plans that looked great on slides and collapsed quickly. The pattern is the same. Teams model TAM from internet research, not SOM from behavior and capacity. Take the steps to force bottom-up math, add market tests before development, and tie decisions to targets, so you'll have numbers you can work with. Confidence goes up, revenue goes up, and hiring decisions make sense. Experimenting and testing are forever skills. Many companies lose that muscle memory as they grow. Never stop measuring your ICP and potentially new ICPs.

Putting It All Together

SOM sets the tone for everything you do. It tells you which channels to test, which sales model to run, and how fast to hire. It also sets the targets in Chapter 2 and the ICP focus in Chapter 4. If your GTM maturity criteria slip, revisit SOM before you add outreach volume or headcount. You do not sell to TAM. You sell to real customers under real constraints.

Companion Resources

Worksheets WS06, WS07, and WS08 support Chapter 3. They cover TAM–SAM–SOM, bottom-up reach, and account capacity modeling.

Download templates, worksheets, checklists, GTM-focused AI bots, and planning tools referenced in the book to help

you apply the content directly to your go-to-market. The materials are updated periodically. Register for updates on our website's Resources page.

The Resources are available by using the QR code in the Companion Resources section at the beginning of the book.

4

YOUR ICP ISN'T WHO
YOU THINK IT IS

"Who is the customer? What does he value?"
— *Peter Drucker*

Insight

Most teams create an ICP on a slide deck and then try to make it exist. But your ICP reveals itself by its behavior. It lives where five forces intersect:

Need: the customer has an urgent problem your product solves.

Fit: the solution aligns with their use case, culture, and tech stack.

Access: you can reliably reach the buyer through outbound, inbound, or referrals.

Economics: they can afford your price and see a clear ROI.

Velocity: the deal moves fast through the funnel: low friction, high intent.

Start with the customers who already buy fast, pay full price, and stay. Those are your radical lovers. They tell you what the market wants, not what you wish it wanted.

Practical Walkthrough

Turn ICP from a guess into a working profile you can sell to this quarter.

SEARCH FOR EVIDENCE

List your last 25 wins and 25 losses. Quantify the sales cycle, discount, ACV, referral, and expansion within 90 days for those 50 deals.

Sort by velocity and price integrity (no discounts). Identify the fastest full price wins. That is your starting set.

RUN 10 INTERVIEWS IN 10 DAYS

Choose six buyers from the starting set, three power users, and three borderline ICP.

Ask for facts, not opinions. Use a script similar to the one below; record and label answers.

BUILD A BEHAVIORAL ICP

Determine traits and triggers. Traits describe the company and role. Triggers explain why they act now.

Add access and economics. Can you reach them this month? Can they pay your price and get ROI?

Add velocity. Do they close fast?

DEFINE THE ANTI-ICP

Name the ICP or persona that slows you down or drags the price down.

Document the red flags you will avoid. Don't waste time with resourcing-draining leads that drag down your GTM.

If you believe strongly in the potential of the resource-draining segment, choose one to nurture, but don't bet your entire business on winning that customer.

OPERATIONALIZE

Build:

Targeting rules for lists and lead-routing.

Messaging that uses the buyer's language.

Proof of value you will bring to the first call.

SLAs for speed to first touch.

Keep or kill thresholds for reply, meeting, and win rates.

Review every two weeks:

If reply quality or win rate slips for two weeks, revisit triggers.

Check for missing proof of value or the wrong buyer if the sales cycle increases.

Example Interview Script

1. What makes this solution a priority now?
2. What did you try before? What worked? What did not?
3. Describe your pain in your own words.
4. Who else feels this pain? Who pays for it?
5. What would a good outcome look like in one month and six months?
6. What must this solution integrate with on day one?
7. What must be true for you to buy?
8. What risks make you hesitate to buy?
9. What proof of value would reduce that risk?
10. If you could keep one feature and lose the rest, what would you keep and why?
11. Who else on your team should we talk to?

Example Behavioral ICP Grid

Fill one column per segment.

	ICP 1	ICP 2
Trait	Head of RevOps, 200 to 800 FTE SaaS	Director of Ops, PE-backed portfolio
Trigger	New CRM migration or RevOps hire in last 90 days	New buy-and-build mandate announced
Access plan	Outbound to RevOps + recent job change list	Partner intro + targeted outbound
Proof of value	Two examples of risk removal and an ROI one-pager	Capacity model and 30-day MAP
Economics	ASP 15k to 40k, discount under 10%	ASP 30k to 80k, budget owner in finance
Velocity signal	First meeting inside 7 days, close under 45 days	Pilot in 30 days, expansion by day 90

Anti-ICP Red Flags

- No named owner of the problem.
- Procurement before a champion.
- Wants custom features before a first value step.
- Needs a committee to approve even a pilot.
- Asks for a discount before a discovery call.

Example

In our own targeting, we moved from a vague "midmarket ops" persona to a behaviorally specific ICP: RevOps leaders who had changed jobs in the past 90 days and owned a CRM project. Same industry. New trigger. Our reply rate rose 3×, sales cycle fell 33%, and median discount dropped from 18% to zero. We didn't overhaul the product. We just sold to the buyers who were already leaning forward.

Our Experience

We have retargeted teams away from desired logos to radical lovers and watched deals accelerate. In one case, the founder wanted enterprise brands. The data said 200 to 600 employees, B2B SaaS, with a new RevOps lead. When we followed the data, replies and wins increased. When the team drifted back to their desired logos, the sales cycle increased, and discounts became necessary. The market does not care about your preferences. It responds to its own needs.

Putting It All Together

The ICP drives everything else. It shapes your SOM in Chapter 3, your message in Chapter 5, your channel choices in Chapters 6 and 7, and the steps in your sales process in Chapter 8. Review and refine your ICP monthly until you consistently close deals with the right customers, which we define as GTM Maturity Stage 2: repeatable wins, reliable proof of value, and stable buyer behavior. Once that holds, shift to quarterly reviews. If your win rate drops or sales cycle slows, revisit your ICP and proof of value before adding volume.

Companion Resources

Worksheets WS03, WS08, and WS09 support Chapter 4. They help you define ICP, map behavioral triggers, and flag anti-ICP segments.

Download templates, worksheets, checklists, GTM-focused AI bots, and planning tools referenced in the book to help you apply the content directly to your go-to-market. The materials are updated periodically. Register for updates on our website's Resources page.

The Resources are available by using the QR code in the Companion Resources section at the beginning of the book.

SECTION II

CRAFTING YOUR GTM ENGINE

5

CLEAR IS BETTER THAN CLEVER

"When you advertise fire extinguishers, open with the fire."
— *David Ogilvy*

Insight

Buyers respond to their words, not yours. Messaging that wins is specific to one pain, one promise, and a proof of value that removes risk. If your message only works when the founder says it, it is not a message. It is a relationship, and it is not scalable. The job is to earn a reply (and win) from the right buyer with the least effort possible.

Practical Walkthrough

Build a simple message system this week.

1. Anchor to ICP and triggers.

Define one real ICP and one trigger. Write the buyer's pain in their words, not your claims.

2. Draft three micro-messages.

Draft three short message variants, each built around a simple flow: pain, promise, and proof of value in no more than three sentences. Use a specific proof point such as a measurable result, a customer quote, a brief case study, or a clear risk reduction.

3. Build a proof of value library.

Collect five core proof items such as two short customer quotes, one before-and-after metric, a two-minute demo clip, and an infosec or compliance one-pager. Store all links where they can be shared easily on the first touch.

4. Run a 10 × 10 message loop.

Run a 10 × 10 message loop to validate your outreach. Send ten messages per variant to a clean list of ten accounts each, then track reply rate, qualified reply rate, meeting rate, and time to first meeting.

10 × 10 Message Loop

Before scaling outreach, test message resonance with a 10 × 10 loop: create 10 distinct message variants and send each to 10 qualified targets (100 total). Monitor opens, replies, and engagement to see what sticks. This is not mass outbound, it's fast, down-and-dirty feedback. Once a winner emerges, scale it and support it with A/B testing.

5. Keep or kill.

- Keep winners only if the qualified reply rate ≥2× the weakest variant and the meeting rate ≥ the target for two consecutive weeks.
- Kill laggards fast; rewrite using the language from the best replies.

6. Fine-tune the opener and the close.

Fine-tune both the opener and the close. Limit subject lines to three variants that highlight a problem, trigger, or ROI cue. Start the message with a 12- to 18-word reflection of the buyer's pain, written in plain language without filler adjectives. End with a single clear decision, such as "Worth a 15-minute compare?" or "Want the 2-minute video?"

7. Short landing page to match.

Create a short landing page that mirrors the message and drives a single action. Keep it to one page, focused on the problem, outcome, proof, and call to action. Remove friction by eliminating navigation links and keeping the layout simple, with a clear calendar link or other relevant CTA.

Example

Target: RevOps leaders who started a new role in the last 90 days and who own a CRM change.

Variant A led with product features. Variant B opened with "Inherited a leaky funnel and a CRM everyone ignores?" and linked a two-minute video clip. In the second email, variant C added a quantified outcome and an IT deployment one-pager.

Assuming these results over two weeks:

- A: 1.4% reply, 0.4% qualified, 0.3% meeting
- B: 4.1% reply, 2.2% qualified, 1.6% meeting
- C: 4.3% reply, 2.8% qualified, 2.1% meeting

Which would you keep? Kill? Rework?

Our Experience

Early in John's career, his outbound messages read like long, formal, and emotionally flat corporate memos. They were overly polite, overexplained, and ultimately ineffective. He was writing to prove credibility, not to earn attention. Worse, his messages failed to speak to the reader's pain and didn't respect their time.

After months of poor response rates, John reframed the problem through a behavioral psychology lens: Every email is fighting to be opened, read, and acted on in a flood of noise. People don't read long emails later; they ignore them forever. But they do scan quickly for one thing: relief.

When the message mentions a problem they already feel, and offers a simple next step that could solve it, they pay attention, especially if that promise comes in the first sentence.

John rebuilt his approach: fewer words, stronger verbs, direct subject lines that mimicked internal messages. He moved the proof earlier, made replies easier, and always led with the buyer's language. The results were fast and measurable: higher reply rates, shorter sales cycles, and better conversations.

The takeaway: clear always beats clever. Show you understand the pain. Earn the reply fast. And never waste a buyer's time.

Putting It All Together

Messaging turns ICP into motion. It feeds the outbound in Chapter 6 and landing pages for inbound and PLG in Chapter 7. It shrinks the proof of value you must deliver in Chapter 8. Treat it like code: build small, test fast, keep what wins, and kill everything else. If the win rate slips, revise and repeat before you add volume.

Companion Resources

Worksheets WS04, WS10, and WS12 support Chapter 5. Use them to draft value propositions, test messages, and to show proof of value.

Download templates, worksheets, checklists, GTM-focused AI bots, and planning tools referenced in the book to help you apply the content directly to your go-to-market. The materials are updated periodically. Register for updates on our website's Resources page.

The Resources are available by using the QR code in the Companion Resources section at the beginning of the book.

6

OUTBOUND ISN'T OPTIONAL

"I would have written a shorter letter, but I didn't have the time."
— *Mark Twain (attr.)*

Insight

Outbound sales is not spray and pray. It is control and learning. Done right, it gives you a fast signal on ICP, message, and pricing, and generates meetings on demand. Done wrong, it burns domains, lists, and goodwill. Start small, run one clean experiment at a time, and keep only what improves reply quality and creates meetings with the right buyers.

Practical Walkthrough

Install a simple outbound program in 30 days.

1. Pick one ICP, one trigger, one offer.

Offer a clear, valuable next step: 20-minute demo, data tear-down, pilot outline, or two-minute video. Write one sentence that ties the trigger to the offer. See ICP and triggers from Chapter 4.

2. Build a clean list.

Identify 100 to 200 accounts that match the ICP with a trigger. Find one to three relevant contacts per account, then validate their email and role. Tag the trigger source so you can report it later.

3. Design one fast multichannel sequence.

Design one fast multichannel sequence with eight to ten touches over fifteen to twenty business days. Use a mix of email, LinkedIn profile views or notes, calls, short videos, and reply bumps. Format each email around pain, promise, and proof, **keeping it to three to five lines.** Test subject lines that reference a problem, trigger, or ROI cue, and open with language that reflects the buyer's own words from Chapter 5.

IMPORTANT: Your goal is to get a meeting, drive a website sign-up, or maybe even a self-service purchase, depending on your sales motion. DO NOT brain dump every detail about your product into your outbound messages. DO NOT talk about product features or technical specifications. You are selling a solution to PAIN, a PROMISE to ease the pain, and PROOF that you can do it. NOBODY cares about the underlying technology, no matter how "cool" you think it is. NOBODY will read your email if they open it and it looks like a Leo Tolstoy novel. **In the first one to two sentences, define the pain, state how you fix the pain, and provide proof you can do it.**

Set clear SLAs and roles for every stage of engagement. Handle all replies within twenty-four hours and follow up on no-shows within one business day. The SDR is responsible for booking and qualifying meetings, while the AE leads discovery and demonstrates proof of value. Log every touch with a clear disposition and next step.

4. Monitor the funnel (track weekly).

Monitor the funnel weekly and track key metrics such as valid contact rate, reply and positive reply rates, meeting and no-show rates, conversion from meeting to opportunity, win rate, and sales cycle length for outbound sources. Add short notes on quality to capture why the positive replies succeeded and the negative replies failed.

5. Run a 10 x 10 message loop.

Run a 10 × 10 message loop to identify what resonates. Test three message variants, each sent to ten accounts under the same list rules. Keep the winning version only if qualified replies are at least twice that of the weakest performer and meet targets for two consecutive sales cycles. Kill underperforming variants quickly and rewrite using the language taken from positive replies.

6. Keep or kill rules: set targets, stick to them.

- Reply rate.
- Positive reply rate.
- Meeting rate with no-show.
- Meeting to opportunity.
- Opportunity to win.

Stop and fix the message or list if the targets are missed.

Always Leave With More Information

Every outbound touchpoint should give you more than you started with. If the decision-maker is out, ask when they're back. If you hit an assistant, ask how to get on the calendar. If you get a "not now," ask what timing works. Outbound dies when founders treat "no" or "away" as the end of the trail. Outbound lives when every step yields the next breadcrumb.

Example

Target: RevOps leaders hired within 90 days and who own a CRM change.

Three variants ran for two weeks to 180 contacts across 120 accounts.

Variant A led with product features. Variant B opened with the trigger and a pain mirror. Variant C reused B's opener and added a two-minute video clip plus a simple MAP in the third email.

Results:

A: reply 1.2%, positive 0.4%, meeting 0.3%

B: reply 3.8%, positive 1.8%, meeting 1.3%

C: reply 4.1%, positive 2.6%, meeting 1.9%, no show 9%

C cleared the targets for two sales cycles, so we kept C, stopped A in the first week, and rewrote B with language gleaned from positive replies.

Our Experience

We have burned through lists by chasing volume before the messaging was sound. We have also tripled meetings by fixing the list quality and opening with a trigger that the buyer already felt. The biggest opening is usually small runs with clear targets and a weekly review. Meeting-to-opportunity increased when we enforced dispositions and next steps, then handoffs stopped leaking.

We also learned the hard way that tone matters as much as tactics. Too many founders sound flat when they talk about their own product. The buyer won't be excited if you don't sound excited. Our rule: whoever does the first outreach must be the company's most animated evangelist.

Every call, email, or social media touchpoint should leave you with more information than you started: the right contact, the timing, and the process to get on the calendar. That discipline turns dead ends into next steps and keeps the pipeline alive.

Putting It All Together

Outbound turns direct ICP messaging into meetings. It feeds Chapter 8 by giving you discovery repetitions that can teach the team not to rely on the founder. It ties in with inbound in Chapter 7 and RevOps in Chapter 9. It conforms to the targets from Chapter 2. If reply quality falls or no-shows spike, fix the message and lead list before adding volume or headcount.

And don't let conversations die. Always follow up a voice-mail with an email and push for the next step: a better email, a meeting booking, or the decision-maker's name. That persistence compounds, just like list quality and messaging.

Companion Resources

Worksheets WS13, WS14, and WS16 support Chapter 6. They cover target lists, sequence design, and outbound SLAs.

Download templates, worksheets, checklists, GTM-focused AI bots, and planning tools referenced in the book to help you apply the content directly to your go-to-market. The materials are updated periodically. Register for updates on our website's Resources page.

The Resources are available by using the QR code in the Companion Resources section at the beginning of the book.

7

INBOUND AND PRODUCT LED GROWTH

"The way to get product-market fit is to build a product so good people spontaneously tell others to use it."
— *Paul Buchheit*

Insight

Inbound and PLG are not free growth; they are systems that either compound or leak. Most teams lose high-intent demand due to slow responses, bad routing, and unclear ownership. PLG stalls when activation is fuzzy, and no one helps the user over the first-value hump. Treat inbound and PLG like a pipeline you can see, measure, and improve. Add a Sales-Assist motion where humans accelerate value.

Practical Walkthrough

Create an inbound and PLG engine that does not leak.

1. Define entry points and intent.

Define your entry points and intent signals. List every path a prospect can take, such as demo or contact forms, pricing page clicks, documentation signups, product trials, and chat interactions. Tag each entry point with an intent level: low, medium, or high. Create clear working definitions for how each level is handled.

Set working definitions:

MQL = "meets ICP traits and shows medium intent."

PQL = "hits an in-product event that predicts an ICP revenue event."

2. Capture cleanly.

Capture data cleanly. Use short forms that include only name, work email, company, and role, adding progressive profiling later. Deduplicate and enrich records after capture rather than before. Always track the source with UTM parameters, referrer, and campaign details.

3. Route with SLAs.

Route leads with clear SLAs. Respond to inbound leads within five minutes during business hours and within sixty minutes after hours. Handoff product-qualified leads to a human within two hours or less. Base routing logic on ICP fit, intent class, territory, and segment, and always assign a backup owner with an overflow rule.

4. Install a Sales-Assist play.

Install a Sales-Assist play that activates when key triggers occur, such as reaching an activation milestone, stalling at a setup step, visiting the pricing page twice, adding three or more seats, or inviting a manager. When triggered, review the user's setup, show the fastest path to value, offer a fifteen-minute working session, and propose the smallest paid plan that fits. Conclude with a short mutual plan that lists dates, owners, and exit criteria for a successful trial.

5. Define activation and first value.

Define what 'activation' and 'first value' mean for your ICP. Identify one to three key events that show a user has reached first value, such as connecting a data source, creating the first report, or inviting two teammates. Make this progress visible to both the user and your team. Create an onboarding checklist designed to help every new user reach first value within seven days.

6. Tighten the product-to-human loop.

Tighten the loop between product and human touchpoints. Utilize in-app cues, such as checklists, tools, tips, and a help button, that route high-intent users to Sales Assist. Send four to six email drips tied to activation steps rather than a fixed schedule. Include a calendar link in the second and third messages, keeping each ask small and specific.

7. Remove predictable friction.

Remove predictable friction from the user experience. Create a clear pricing page with one recommended plan tailored to your ICP. Make it simple to upgrade from a free or trial account, and remove any gated content once a user has signed up.

8. Monitor the funnel.

Monitor the funnel across the entire journey: track conversion from visitor to lead, MQL rate, and PQL rate. Measure time to first touch, activation rate, and time to first value. Watch trial-to-paid conversion, paid plan mix, and expansion by day 30 and day 90.

Set targets per segment. For example, first touch ≤5 minutes, activation ≤7 days, trial to paid ≥20% for SMB and ≥10% for mid-market, expansion by day 90 ≥15%.

9. Run experiments for two sales cycles.

Run experiments over two complete sales cycles to measure real impact. Reduce the number of form fields from six to four and track changes in MQL quality. On the pricing page, test the effect of adding or adjusting a recommended plan label. Compare trial lengths of 14 versus 21 days to determine which drives stronger conversion and engagement.

Sales-Assist: When activating users for the next event or milestone, add a link that allows them to book a short working session (e.g., a 1:1 help call or demo). Then, track whether this addition actually improves your key activation metric.

Example

A team with solid inbound saw trial signups but weak conversion. PQLs sat unworked overnight, and users stalled before connecting their first data source. The team set a two-hour SLA for human contact for PQLs, defined three activation events, added a 15-minute Sales-Assist working session trigger at event 2, and simplified the pricing page to one recommended plan. Activation to paid rose 18%, and time to first value dropped from 12 days to 6. Expansion by day 90 improved by 11%.

Our Experience

We find "hot leads" dying in forms, chatbots, and trials. Conversions are moved once short response rates are enforced (e.g., on demo requests), and two-hour human coverage on PQLs, along with a clear activation definition, is sought. The most significant shift came from adding real sales assistance. Short human interventions at the right moments beat longer free trials and more emails.

Putting It All Together

Inbound and PLG ride on the RevOps backbone you built in earlier chapters. ICP and messaging shape your pages and prompts. RevOps supplies routing and SLAs. The first sales process defines proof of value and sets the next steps. Reporting makes the leaks visible. If trial-to-paid stalls, do not add more content. Fix capture, routing, activation, and sales assistance first.

Companion Resources

Worksheets WS10, WS11, and WS12 support Chapter 7. Use them to plan inbound content, map PLG signals, and run small A/B tests.

Download templates, worksheets, checklists, GTM-focused AI bots, and planning tools referenced in the book to help you apply the content directly to your go-to-market. The materials are updated periodically. Register for updates on our website's Resources page.

The Resources are available by using the QR code in the Companion Resources section at the beginning of the book.

8

ALIGNMENT IS THE
GTM FORCE MULTIPLIER

"Plans are worthless, but planning is everything."
— *Dwight D. Eisenhower*

Insight

Misalignment makes the entire team look busy, but it sells nothing. When Marketing chases volume, Sales improvises, Product ships features that don't win deals, and Customer Success fights fires, you're paying four teams to pull in different directions. Alignment is a system, not a slogan. It means one customer journey, shared definitions, clear decision rights, clean handoffs, and a cadence that turns learning into action every week.

It is worth distinguishing the customer journey from the sales process. The customer journey is how the buyer advances toward a decision, whereas the sales process is how your team advances the deal. They have to agree. Design your sales process to follow the buyer's journey, optimizing each touchpoint to reduce friction, shorten the sales cycle, and increase potential deal value.

The customer journey is like water. Customers take the path of least resistance. Your job is to reduce friction. Don't let the path of least resistance lead to your competitors.

Practical Walkthrough

Install an alignment system that removes friction and accelerates decisions.

1. Map one customer journey.

Map a single, complete customer journey. Define the key stages such as Aware, Engage, Qualify, Proof, Align, Commit, Onboard, and Expand. For each stage, document the owner, entry signal, exit criteria, required artifacts such as emails, MAPs, or order forms, and the SLA that governs movement to the next stage.

2. Standardize definitions.

Standardize your key definitions so every team speaks the same language. Define lead, MQL, SQL, PQL, and each opportunity stage. Clarify your ICP and the triggers that signal readiness. Write one-line definitions for "Qualified" and "Success Criteria" in the proof of value, and record them in both the CRM and the playbook.

3. Set RACI decision rights.

Set clear RACI decision rights to eliminate overlap and confusion. Define who proposes, approves, and informs for pricing and discount rules, channel additions or cuts, segment updates, ICP edits, and tooling or schema changes. If two people believe they own a decision, no one owns it. Assign explicit ownership for every action.

4. Establish a cross-functional pipeline review.

Establish a cross-functional pipeline review cadence that keeps every team aligned. On Monday, hold a 30-minute GTM stand-up to identify a bottleneck, assign an owner, deploy one solution, and set a rollback date. Midweek, review the pipeline for stage accuracy, time in stage, and conversion waterfalls. On Friday, run a short product and GTM sync to share what was learned, what will be tested next, and what will be implemented to support it. Keep every meeting brief and data-driven, and avoid slides when the CRM can show the same information.

5. Enforce clean handoffs.

Enforce clean handoffs across teams to maintain momentum and clarity. Every handoff must include notes, the next step, and event dates recorded in the CRM. When a deal closes, Sales provides Customer Success with a clear first-value plan that must be confirmed within twenty-four hours. Pause any deal that lacks an agreed-upon MAP after discovery.

6. Build an enablement pack that matches the customer journey.

Build an enablement pack that mirrors the customer journey. Include buyer-language messaging lines, an objection bank with proof of value, a two-minute demo video clip, a compliance one-pager, pricing details, give–get rules, and a MAP template. Store the pack where AEs can access it instantly without searching; maintain a single source of truth for all materials.

7. Install a shared dashboard.

Install a shared dashboard that gives every function the same view of performance. Include SQLs per month, win rate, median sales cycle, median discount, and new ARR. Display stage conversion waterfalls, source mix across outbound, inbound, PLG, and partner channels, and SLA hit rates for inbound and PQL responses. Ensure all numbers match across teams, and if any metric still requires a spreadsheet merge, automate it to maintain a single source of truth.

8. Run a friction log.

Run a friction log to identify and remove barriers that slow deals. List the top five frictions by impact, such as legal, infosec, procurement, data access, or integrations. Assign an owner, solution, and due date for each. Start by removing the frictions you control directly, including forms, SLAs, proof of value steps, paperwork, and terms and conditions.

9. Light governance.

Apply light governance to keep systems stable without slowing execution. Manage change control for pick lists, routing, and automations. Review quarterly key definitions, discount bands, and the data or AI policy. Assign clear ownership for both the CRM schema and the company glossary.

Example

A team with strong inbound looked healthy until deals died at legal and IT. Marketing measured MQLs, Sales counted late-stage pipeline, and Product shipped features that did not address pain. Customer Success met customers cold. The team mapped one journey, moved legal and IT into the align stage with exit criteria, and required a MAP within 48 hours of discovery. Within six weeks, stage accuracy hit 96%, the median sales cycle fell by 18 days, and discounts dropped below 12% without changing the price.

Our Experience

We have seen leaders try to repair misalignment with new tools. Tools amplify whatever system you already run. The noise stops when you force one journey, define decision rights, and establish a weekly cadence. Moving ownership of "pricing exceptions" from an ad hoc email to a named approver cut discount variance in half and raised the win rate by 5% in one client company. The fix was a process, not software.

Putting It All Together

Alignment helps every chapter before this one work simultaneously. ICP and messaging shape the journey. Outbound, inbound, and PLG plug into shared SLAs. The first sales process establishes proof of value and handoffs. RevOps keeps definitions and data honest. Pricing sits inside clear give-get rules.

Next, Chapter 9 turns pricing into a GTM strategy (not a finance exercise).

Companion Resources

WS22, WS20, and WS18 support Chapter 8. They help you map the sales process, define handoffs, and set MAPs.

Download templates, worksheets, checklists, GTM-focused AI bots, and planning tools referenced in the book to help you apply the content directly to your go-to-market. The materials are updated periodically. Register for updates on our website's Resources page.

The Resources are available by using the QR code in the Companion Resources section at the beginning of the book.

9

PRICING IS STRATEGY, NOT A FINANCE EXERCISE

"Price is what you pay. Value is what you get."
— *Warren Buffett*

Insight

Pricing is not a spreadsheet. It is how you position value, qualify buyers, and control deal quality. Bad pricing creates long sales cycles, heavy discounts, and customer churn. Good pricing clarifies who you serve, what outcome you deliver, and how you get paid. Treat pricing like a product. Form a hypothesis, run fast tests on small cohorts, keep what wins, and lock it in long enough to prove repeatability. Your customers aren't only buying your product; **they are buying the entire experience with your company, from first contact onwards.**

> **Every touchpoint is a tipping point.**
> **Every person and every process that meets your customer either makes the sale or breaks it.**

Practical Walkthrough

Build a pricing strategy you can test and teach in one sales cycle.

I. Choose a value metric and an anchor.

Pick one metric your ICP already uses to measure value: seats, active users, volume of jobs, revenue influenced, or records processed. Anchor your story to a clear outcome: save X hours, capture Y revenue, reduce Z risk. The price follows the outcome.

2. Draft a simple product package ladder.

Spell out what "Good," "Better," and "Best" offerings tiers mean for your ideal customer types. Each tier should have one main outcome, two or three key features, and one clear usage limit. Make sure customers can easily see and understand those limits. Don't hide them in the fine print.

3. Set list price, then define discount guardrails.

Publish list prices in your enablement pack. Define a discount band by segment with approvals. Use give-get rules:

every discount trades for a term, case study, multiyear, or prepay. There are no silent discounts.

4. Design two pricing experiments.

Example: [1] tier names and limits, and [2] annual-only vs. monthly with a surcharge.

Run one cohort at a time for a sales cycle. Hold message, ICP, and channel constant. Define pass-fail thresholds such as win rate, median discount, sales cycle days, ACV, and payback.

5. Install a basic deal desk.

Who can approve which discount levels? Implement the 'rule of three.' If you need approval from more than three people within your company to complete a deal, it is probably too complicated or off-model. Either simplify it so it fits your usual process, or walk away.

Keep a change log of sales contracts. Watch out for discount creep or other anomalies.

––––––

HOW TO USE THE RULE OF THREE

Lay out the standard path in a MAP: who signs, which proofs of value, and standard terms. If the deal will need more than three special approvals or exceptions across both sides (you and your customer) to get the deal done, treat it as a bad fit.

First, reframe to the standard path or price the complexity with a clear give-get. If you cannot reduce the count to three or fewer special considerations, walk away and protect focus.

The point here is to avoid wasting time on complex deals that require one or both sides to jump through hoops. No matter how badly you want the logo, or how big the deal value is, or whatever dreams you or one of your AEs has about this account, chasing a bad fit creates risk. The risk that the deal will drag on and never close. The risk that if the deal does close, the customer will churn quickly. The risk that the special requirements create a massive burden on your company's resources to service the customer. For small start-ups, a poorly fitting deal could put you out of business.

Heuristics:

- SMB or midmarket: needing one to two approvals is normal.
- Enterprise: up to three is fine if pre-mapped with a real champion.
- More than three usually means weak fit, weak champion, or poor segmentation.

––––––––

6. Train the price conversation.

You should lead with outcome and proof, not a price sheet. Handle "too expensive" with a value and scope comparison, not a reflex discount. Offer the smallest paid plan that fits. Expansion is part of the story. Get your foot in the door and grow the account.

7. Develop the pricing funnel.

Track deals by source and segment, including win rate, median discount, sales cycle days, ACV, churn signals at 90 days, and CAC payback. If the win rate rises, but so do the discount and sales cycle, you are buying deals. Fix positioning and proof of value before you add volume.

8. Keep or kill rules.

Keep any change that improves win rate and ACV without lifting the discount or sales cycle. Kill any change that inflates discounts or extends the sales cycle without expanding ACV or payback. Lock pricing for at least one complete sales cycle before the next test.

Example

A midmarket workflow SaaS business sold "custom quotes" and bled discounts. The team built a three-tier ladder, published a list internally, and installed give-get rules. They ran a two-week test comparing annual-only plans with a 10% prepay incentive to monthly plans with a surcharge. For the annual cohort, the median discount dropped from 19% to 11%, the sales cycle fell by 9 days, and the ACV rose 14%. The company kept annual-first for the midmarket and retained monthly for SMB only.

Our Experience

We have watched founders treat pricing like a secret and complain about long sales cycles and heavy discounts. When we made price part of the story, connected it to a value metric, and enforced give-get rules, deals moved faster, and discount variance shrank. In one client company, the only change was requiring a discount greater than 10% to trade for a one-year prepay. The median discount fell, and cash collected improved in the same quarter.

Putting It All Together

Pricing touches everything. ICP and SOM define who can and will pay. Messaging sets the value story, and outbound and inbound set expectations. The first sales process provides the proof of value that justifies the price. RevOps enforces approvals and logs exceptions. Reporting surfaces payback and discount creep. If deals stall on price, fix the proof of value and messaging first. Price is a strategy, not an excuse to give out discounts.

Companion Resources

Worksheet WS17 supports Chapter 9. It is a RevOps backbone and data hygiene audit to stabilize your pipeline.

Download templates, worksheets, checklists, GTM-focused AI bots, and planning tools referenced in the book to help you apply the content directly to your go-to-market. The materials are updated periodically. Register for updates on our website's Resources page.

The Resources are available by using the QR code in the Companion Resources section at the beginning of the book.

SECTION III

SELLING AND SCALING

10

THE FIRST SALES PROCESS

"An organization's ability to learn, and translate that learning into action rapidly, is the ultimate competitive advantage."
— *Jack Welch*

Insight

You do not have a process if your process lives in the founder's head. You have founder-dependent stories. Before you hire anyone, design a straightforward path from first touch to closed won that someone else can run. Define the stages, the proof of value, all the stakeholders who must say yes, and the exit criteria for each step. It is not a deal if a deal cannot move along this path. It is a time-suck that will end in no deal.

Practical Walkthrough

Build a teachable sales process in one week.

1. Draw one path to a win, for example:

Stages: Triage → Discovery → Prove → Align → Commit → Win or Loss.

Draw a clear path to a win and define each stage in the journey. Use stages such as Triage, Discovery, Prove, Align, Commit, and Win or Loss. In Triage, confirm that the deal matches your ICP and has an active trigger. During Discovery, validate the problem, impact, priority, stakeholders, timeline, and risks. In Prove, deliver the minimum proof of value needed to remove risk. In Align, finalize pricing, paperwork, and the plan with the buying committee. During Commit, secure agreement on timing and terms. Whether you win or lose, record the outcome and the reasons.

2. Set entry and exit criteria and document the facts needed to move a deal forward in the pipeline. No opinions. For example:

Set clear entry and exit criteria for every stage in the pipeline and document only the facts needed to move a deal forward. Avoid opinions or assumptions. For example, the Discovery exit should include a named champion, a problem defined in the buyer's words, a value metric, the decision process, required integrations, timing, and a scheduled next meeting. The Proof exit should confirm that the proof of value is complete, the success criteria are met, written feedback is captured, and legal or IT are engaged if needed. The Align exit should include a written agreement on price, an accepted MAP, redlines in progress, and a confirmed target close date.

3. Design the proof of value. Pick one format for this stage: demo, pilot, or timeboxed trial.

Design a focused proof of value for this stage using a single format such as a demo, pilot, or timeboxed trial. Define the success criteria in one clear sentence and keep the steps and owners to a minimum. Conclude with a summary email that confirms what proof was demonstrated and aligns both teams on the outcome.

4. Use a MAP.

Use a Mutual Action Plan (MAP) to keep deals organized and accountable. Include decision steps, owners on both sides, key dates, and exit criteria. Send the MAP on the first day of discovery and review it at the start of every call. Any deal without an active MAP should be paused or marked closed lost.

5. Handoffs and roles.

Define clear handoffs and roles to maintain consistency across the funnel. The SDR qualifies and books meetings, the AE leads discovery and delivers proof of value, and the CSM joins only for the defined stage. Each handoff must include notes, the next step, and a date recorded in the CRM. If a handoff fails twice in a row, correct the process before increasing volume.

6. Instrumentation and hygiene.

Set up the fields and tracking needed to measure pipeline health. Capture ICP fit, trigger, champion, stakeholders, value metric, next step with date, stage-specific exit fields, and accurate loss reason. Apply time-in-stage thresholds, and if a deal passes the limit without a next step, park it. Record all calls and tag who attended, such as technical, legal, buyer, or IT.

7. Asset pack examples.

Build an asset pack that supports every stage of the deal. Include a two-minute demo video clip, a one-page summary of pricing and give–get rules, an infosec overview, an objection and answer list, and three email templates for follow-up and MAP confirmation.

8. Weekly pipeline triage & hygiene.

Maintain weekly pipeline hygiene to keep data accurate and decisions reliable. Ensure every active deal includes a defined next step, and keep stage accuracy above ninety-five percent. Track median sales cycle days by stage and win rate by source. Implement one improvement each week, assign an owner, and set a rollback date to measure its impact.

9. Founder-less test.

Teach the process to one AE. For two sales cycles, the AE must reach at least 80% of the founder's win rate and SLA adherence without the founder in the room. Don't hire more AEs until the AEs you have can close alone.

Example

A team selling workflow software treated every trial as a custom project. Legal and IT showed up at the end, and deals were restarted. The team mapped one path and moved legal and IT to the Align stage with clear exit criteria. They then replaced open trials with a 14-day pilot with a checklist, one value metric, and a MAP. Time in stage fell by 30 days, and the win rate rose from 14% to 24% without adding headcount.

Our Experience

We have seen founders try to scale their personality. It does not scale. The turning point comes when the process becomes the product. Adding a nonnegotiable rule changed the outcome in one case: every late-stage deal needed a MAP with names and dates. Deals without it were parked. Within four weeks, forecast accuracy improved, and legal stopped torpedoing the last mile.

Putting It All Together

This chapter turns your ICP and messaging into a path anyone on your team can follow. The process you document here becomes the backbone for RevOps in Chapter 9, the hiring target in Chapter 12, and the stage gates in Chapter 2. When win rate slips or sales cycle days rise, fix your exit criteria and proof of value before adding more leads or AEs.

Companion Resources

WS15, WS14, WS16, and WS20 support Chapter 10. They align teams on goals, cadences, SLAs, and operating rhythms.

Download templates, worksheets, checklists, GTM-focused AI bots, and planning tools referenced in the book to help you apply the content directly to your go-to-market. The materials are updated periodically. Register for updates on our website's Resources page.

The Resources are available by using the QR code in the Companion Resources section at the beginning of the book.

11

HOW REVOPS SHOULD LOOK

"The goal is to turn data into information, and information into insight."
— *Carly Fiorina*

Insight

RevOps is not reports. It is the operating system that keeps Marketing, Sales, Product, and Customer Success running on the same truth. When RevOps is established, you see the pipeline, trust the numbers, and confidently adjust to change. When it is theater, you get pretty dashboards, stale records, and illogical explanations. The RevOps backbone is a clean data model with clear lifecycle definitions and SLAs that people obey and hygiene rules that never bend.

Practical Walkthrough

Install the RevOps backbone in less than 30 days. Do not automate before you can run it manually.

1. Minimum data model.

Create a minimum data model that captures only what is essential to run and measure your funnel. For accounts, record the name, domain, segment, and ICP fit. Label each contact by role, such as champion, technical, legal, or economic buyer. Log all activities, including emails, calls, meetings, and notes with outcomes. For opportunities: track source, stage, owner, amount, products, close date, next step with date, and loss reason. Maintain a simple product table listing features, prices, terms, and discount rules.

2. Lifecycle and stage definitions.

Define each stage of the customer and deal lifecycle. A lead is any person captured. An MQL meets ICP traits and shows medium intent. An SQL is an accepted opportunity with a meeting booked. Opportunity stages include Triage, Discovery, Proof of Value, Align, Commit, and Win or Loss. A PQL shows in-product behavior that predicts revenue for your ICP. Write one sentence to describe each stage and add entry and exit criteria as required fields in the CRM.

3. Examples of SLAs that stop leaks.

Set SLAs that prevent leaks and maintain accountability throughout the funnel. Respond to inbound leads within five minutes, and schedule the first meeting for each MQL within two business days. Reply to all positive responses within twenty-four hours, and update opportunities within twenty-four hours of every call. Record the reason for closed lost deals at the moment they close.

4. Hygiene rules that never bend.

Establish hygiene rules that are never broken. Every active deal must include a next step with a date. The stage must reflect reality and move forward only when the exit criteria are complete. Ensure there are no orphan accounts or contacts. Every record must have an owner. Use a pick list for loss reasons instead of free text so data supports decisions. Park any deal that exceeds its time-in-stage threshold.

5. Routing that respects ICP and intent.

Set up routing rules that respect both ICP fit and intent. Route leads based on ICP match, intent level, and segment, and always assign a backup owner with an overflow rule to prevent delays. For PQLs, ensure a human follows up within two hours.

6. Enablement and give-get rules.

Create enablement and give–get rules that support consistent execution. Define clear price and package guardrails, set discount bands with required approvals, and ensure the MAP template is available for AEs from day one. Maintain an objection bank that links each objection to its supporting proof of value.

7. Weekly pipeline review.

Hold a weekly pipeline review to keep data clean and the process improving. Begin with a twenty-minute hygiene block to fix missing next steps and incorrect stages. Review stage accuracy, time in stage, conversion rates, and win rate by source. Log one change to test in the coming week, assign an owner, and set a rollback date. Finish with a review of closed lost deals, identify the top three reasons, and take action to remove the frictions within your control.

8. A CEO dashboard that fits on one page.

Build a CEO dashboard that fits on one page and shows only the metrics that matter. Include SQLs per month, win rate, median sales cycle, median discount, and ARR added. Add a stage conversion waterfall, source mix across outbound, inbound, PLG, and partner, and SLA hit rates for inbound first touch and PQL to human. All metrics must come directly from the CRM. If a number still requires a spreadsheet, fix the data, automate the calculation, and add a data merge.

9. Data warehouse readiness test.

You likely don't need a data warehouse if a clean CRM answers 90% of questions. You need a data warehouse when any of these are true: five or more data sources to reconcile; recurring CSV merges; multiproduct or multilocation complexity; reverse ETL needed to push truth back into tools; or more than two hours to refresh the executive dashboard.

If ready, the minimal stack is a data warehouse, ELT, modeling layer, reverse ETL, and access governance. Start small.

10. Change control and governance.

Establish precise change control and governance to keep systems stable. Assign an owner for the schema, pick lists, automations, and routing. Version all changes and test them in a sandbox before deployment. Maintain a living glossary that documents every definition, and review the data policy and access permissions quarterly.

Example

A team had three versions of the truth: Marketing reported leads, Sales reported pipeline, and Finance reported bookings. None matched. The team installed the minimum data model, defined lifecycle states with required fields, set a 5-minute inbound SLA, and enforced a next step on every active deal. Stage accuracy moved above 95% in less than a month. The median sales cycle fell by 14 days. The CEO dashboard no longer needed manual spreadsheets.

Our Experience

We have walked into CRMs that looked busy and produced nothing. After the team set SLAs, added required exit fields, and enforced a weekly hygiene meeting, reply quality rose, meetings turned into opportunities, and stage creep stopped. The dramatic part was not a new tool. It was the moment the team realized the system was telling them what to do next, and they could trust it.

Putting It All Together

RevOps is the backbone of everything you built in Chapters 1 through 10. It makes your stage gates in Chapter 2 visible. It makes ICP and messaging measurable. It powers your outbound in Chapter 6 and your inbound and PLG in Chapter 7. It also sets you up for the reporting you will deliver in Chapter 18. If you cannot refresh your CEO dashboard in under an hour from the CRM, fix the backbone before you add tools.

Companion Resources

WS18, WS19, WS20, and WS21 support Chapter 11. They cover pricing and packaging, discount give-get, approvals, and deal desk rules.

Download templates, worksheets, checklists, GTM-focused AI bots, and planning tools referenced in the book to help you apply the content directly to your go-to-market. The

materials are updated periodically. Register for updates on our website's Resources page.

The Resources are available by using the QR code in the Companion Resources section at the beginning of the book.

12

HIRING YOUR FIRST SALESPEOPLE

"First who, then what."
— *Jim Collins*

Insight

The first sales hires determine whether your company builds a repeatable go-to-market system or burns cash in a guessing game. Most founders hire too early, too senior, or too vaguely. A great salesperson can't rescue a broken system; they simply reveal whether your sales motion is teachable.

Hiring your first AEs or Sales Development Representatives (SDRs) isn't a bet on personality; it's a test of process integrity. Your goal is to prove that someone who isn't the founder can sell. Until that happens, you don't have product-market fit; you have founder-market fit.

As Bray and Sorey emphasize in their *The Sales Enablement Playbook*, hiring success depends on structured enablement, not intuition. Top companies "get new hires selling on Day 2" because they treat onboarding as a revenue event, not an HR ceremony (Bray & Sorey, 2017, ch. 3). Your first salespeople are proof points for whether your process, training, and culture are scalable.

Practical Walkthrough

1. Confirm you're ready to hire.

Confirm that your company is truly ready to hire. Do not hire salespeople to fix demand; they will drown. Hire them to scale what already works. You are ready when at least 70% of recent wins have followed a documented process for two or more sales cycles, when you can explain win and loss reasons without guessing, and when one offer, proof of value, and pricing model consistently convert. If these conditions are not met, the bottleneck is your system, not your AEs.

2. Choose the right first role.

Choose the right first sales role by matching the hire to your biggest bottleneck. Founders often default to hiring closers when they actually need builders. Hire a founder-friendly AE when meetings are consistent and you need someone who can manage discovery, proof of value, and negotiation without founder support. Add an SDR when pipeline creation is weak, but your message performs well once conversations begin. Introduce a Sales-Assist role when product-led growth is strong but users stall before the first paid conversion. Never start with a VP of Sales; that remains the founder's job until the motion is proven (Bray & Sorey, 2017, ch. 3).

3. Design the sales enablement system before day one.

Borrowing from Bray and Sorey's approach, readiness means your onboarding playbook is prebuilt, not improvised. New reps should sell with supervision by day one. Include:

- A clear sales process (stages, exit criteria, SLAs).
- A "whiteboard demo" certification—can they explain value without the slides?
- Standardized talk tracks and objection banks.
- 10 recorded calls labeled by stage.
- A 30–60–90-day plan with specific outcomes.

The onboarding target is to cut ramp time by half compared to peers (Bray & Sorey, 2017, ch. 3).

4. Use a structured hiring scorecard.

Ensure your first sales hires succeed on purpose, not by luck or intuition. The scorecard should evaluate what matters. Keep it short and measurable. Example:

Competency	Weight	Examples
Runs a clean discovery	20%	Can repeat your discovery flow live and write a short summary with next steps
Understands how to use proof of value	20%	Can scope a pilot, set exit criteria, and write a concise recap email
Uses buyer's language for follow-ups	15%	Two short follow-ups that mirror ICP words and link proof of value
Forecasts and updates accurately	15%	Knows stage definitions, enters next steps and dates without prompts
Coachability	15%	Incorporates feedback between interviews and in role play
Domain and ICP fluency	15%	Learns your buyer's world and common objections

5. Build a culture of deliberate practice.

Sales enablement is not training; it's continuous iteration. Bray and Sorey (2017, ch. 4) argue that top performers improve faster because they practice deliberately and receive specific feedback. Mirroring their approach:

- Annotate discovery notes: highlight missed questions or weak summaries.
- Simulate deals: use structured role-plays with real buyer personas.

- Let SDRs close small deals to accelerate learning and strengthen career paths.
- Coach visibly: senior reps model live calls and share recordings weekly.

6. Compensate for stage, not hope.

Compensate for your company's current GTM stage, not for hope. Design ramp and compensation plans that reflect your maturity and data, not future assumptions. Set AE on-target earnings based on a realistic eighty percent attainment. Structure SDR compensation as base pay plus variable tied to qualified meetings that convert into pipeline. Create a ramp plan where month one focuses on shadowing and certification, month two on pipeline building, and month three on closing first deals. Pay commissions only on clean deals that adhere to CRM processes and hygiene standards.

Our Experience

We've walked into too many teams hoping a "name brand" AE would save them. It never works. Big résumés can't compensate for broken RevOps foundations. Performance stabilizes only when hiring runs through scorecards, onboarding starts on day two, and managers inspect behavior weekly.

We've seen founders hire SDRs when they needed Sales-Assist reps, or bring in VPs before proving any motion. In every case, results improved the moment the team simplified: one motion, one proof step, one playbook, one change per week. Enablement always beats charisma.

Putting It All Together

Hiring salespeople is an operational test, not a personality test. Bray and Sorey (2017, chs. 3–4) remind us that effective enablement creates consistency long before culture can. The goal is to shorten ramp time, make success teachable, and create a bench of sellers who think like operators.

Every chapter before this feeds the outcome: ICPs and messaging define conversations; outbound and inbound drive meetings; the process provides the map; RevOps enforces SLAs and hygiene; pricing gives boundaries. When new hires struggle, fix the system first. If the system holds and they still miss, use the scorecard and ramp plan to decide quickly and move on without drama.

Sales success at scale isn't luck. It's enablement, discipline, and design.

Companion Resources

WS23 and WS24 support Chapter 12. Use them for role scorecards, interview loops, and 30–60–90 day onboarding plans.

Download templates, worksheets, checklists, GTM-focused AI bots, and planning tools referenced in the book to help you apply the content directly to your go-to-market. The

materials are updated periodically. Register for updates on our website's Resources page.

The Resources are available by using the QR code in the Companion Resources section at the beginning of the book.

13

CUSTOMER FEEDBACK IS A DEADLY WEAPON

"There is only one boss: the customer."
— *Sam Walton*

Insight

Most teams collect feedback and change nothing. Notes pile up in tickets, call recordings, and surveys, while the loudest voice chases the product roadmap. Feedback is only valuable when it becomes a decision. Build a simple loop that captures facts, tags them the same way every time, ranks by impact, assigns an owner, and closes the loop with the customer and the team.

Practical Walkthrough

Establish a Voice of the Customer (VoC) system in 10 days.

1. List the sources.

List every source of customer insight so you can capture a complete view of feedback. Include sales calls, trials, and pilots; support tickets and chat transcripts; onboarding and QBR notes; NPS, CSAT, CES, and churn interviews; and inputs from community discussions, online reviews, and social channels.

2. Capture in one format. Create a short template inside the CRM or help desk. Required fields:

Capture all customer feedback in one consistent format using a short template inside the CRM or help desk. Include required fields for account and segment, the stage in the journey when the issue appeared, theme, severity tags from pick lists, a running frequency count, and the revenue at risk or potential upside. Add a verbatim buyer quote and a link to proof of value, such as a recording, ticket, or screenshot.

3. Use a shared taxonomy. Start with 8 to 10 tags. Keep them stable for a quarter. Examples:

Use a shared taxonomy to keep feedback organized and comparable across teams. Start with eight to ten tags and keep them consistent for a full quarter. Common tags include product gap, pricing and packaging, UX friction, onboarding and activation, integrations and data, security and compliance, performance and reliability, and buying friction or procurement. Classify each item by severity as either a blocker, material issue, or minor issue.

4. Tag within 24 hours. Whoever touches the customer tags the item before the day ends. A RevOps or CSM owner audits tags twice a week for consistency.

5. Prioritize with a simple score.

Prioritize feedback using a simple scoring model. Rate each item on four scales from one to five: impact on revenue or retention, frequency across accounts, strategic fit with your ICP and stage, and effort to fix. Calculate the priority score as:

(Impact × Frequency × Fit) ÷ Effort

Maintain two backlogs: one for GTM fixes and one for product fixes. Many issues that appear to be product problems are actually GTM problems in message, proof of value, process, or pricing.

6. Close the loop.

Close the loop on every piece of feedback. Together with the customer, explain what action you will take and when you will follow up. With the AE or CSM, provide the asset or answer they can use immediately. Internally, update the talk track, objection bank, and any related website or documentation as needed.

7. Run a weekly VoC review: one page, 30 minutes.

Run a weekly Voice of Customer review as a focused thirty-minute session with a one-page summary. Highlight the top five themes from the week, identify three GTM changes that can be delivered immediately, and select one product issue for triage. Assign owners and due dates for each action, and update the change log to track progress.

8. Install a win and loss program.

Install a structured win and loss program to capture honest feedback and pattern recognition. Conduct five monthly interviews, balancing wins and losses, and use a neutral interviewer when possible. Ask questions that uncover the buyer's journey: what triggered their search and what had to be true to buy, which alternatives they considered, what nearly stopped the deal, what proof of value created confidence or what risk prevented it, and how your price and product compared. Tag each interview using the standard CRM template for consistent analysis.

9. Outcome metrics.

Track outcome metrics to measure the impact of every change. Monitor reply quality on outbound after message updates, trial activation rate after onboarding improvements, win rate, and sales cycle length for fixes addressing buying friction, expansion by day ninety after success changes, and churn reasons by cohort after product or pricing adjustments. Track the percentage of the roadmap influenced by tagged Voice of Customer data.

Apply strict keep-or-kill rules: keep a change only if it improves the target metric for two consecutive sales cycles. Require at least three independent signals across two segments before adding a major product item to the roadmap, and never build for a single large customer unless it matches your ICP and improves the overall score.

Example

Deals kept dying in legal and security. Feedback was scattered and emotional. We tagged every mention under security and compliance, attached the questionnaires, and counted frequency by segment. Within two weeks, we shipped a short infosec one-pager, published a security page, and added a fast questionnaire path. We also wrote a two-line talk track for AEs with a link to the materials. Legal escalations dropped, sales cycle improved by 12 days, and win rate rose four points without building a feature.

Our Experience

We have seen teams run NPS and then argue about its meaning. Once we developed a shared template, a standard tag set, and weekly decisions, minor fixes paid off fast: a clearer pricing page, a shorter trial setup, and a MAP example in the proof of value stage. With one company, a single change to onboarding emails based on 10 similar quotes doubled activation inside two weeks. The work was not a new tool. It was discipline.

Putting It All Together

Customer feedback connects to everything. It sharpens ICP in Chapter 4, rewrites messaging in Chapter 5, removes friction for outbound and inbound in Chapters 6 and 7, and tightens the proof of value step in Chapter 10. RevOps in Chapter 11 keeps the tags and definitions honest. Reporting in Chapter 18 shows which changes paid off. Treat feedback like code: (1) deliver small, (2) measure, and (3) keep only what moves the metric.

Companion Resources

WS25, WS26, and WS35 support Chapter 13. They set up your customer feedback program, interview scripts, and NPS loop.

Download templates, worksheets, checklists, GTM-focused AI bots, and planning tools referenced in the book to help you apply the content directly to your go-to-market. The materials are updated periodically. Register for updates on our website's Resources page.

The Resources are available by using the QR code in the Companion Resources section at the beginning of the book.

14

INVESTORS, BOARD DIRECTORS, AND ADVISORS

"Trust, but verify."
— Ronald Reagan

Insight

You will drown in advice. Some of it is gold, but much of it is noise. The difference is not how famous the source is, but whether the advice maps to your stage, buyer, and numbers, and whether you can test it quickly. Treat outside input like code. Define the problem, select the correct reviewer, run a controlled change, and keep only what moves a metric you care about.

Practical Walkthrough

Install a simple system for getting, testing, and using advice.

1. Define the decision before you ask.

Document the decision in one line and tag it as either reversible or hard-to-unwind. List the metric that should move and by how much. Add the three constraints that must hold.

2. Use RACI.

    R - Who is directly **responsible?**
    A - Who is the final **approver?**
    C - Who must be **consulted?**
    I - Who must be **informed?**

Document this process and point to it when opinions collide.

3. Build a tight advisor bench. Three seats only at first:

Find advisors across disciplines valuable to you, such as an operator coach (CEO, founder) who has run your motion at your ACV, a domain buyer who matches your ICP and knows what your ICP needs and how they buy it, and a utility operator with a RevOps or finance background who can check your math, projections, and assumptions.

Compensation: Typically, this is via a vested equity plan, but it may depend on your stage and preference.

4. Run the advisor fit test.

Context match: Have they built something at your stage?

Proof: Can they show the before and after?

Availability: Will they make time this quarter?

Skin in the game: Aligned comp and clear scope

Conflicts: No competing boards or customers that create risk.

Accept an advisor only if all five are true.

5. Triage any advice with VRMON.

- Verify the premise with your data.
- Relevance to your stage and ICP.
- Mechanism for how it would work here.
- Ownership of the test and the result.
- Next test is defined with a metric and a date.

If you cannot answer VRMON in 10 minutes, park the advice.

6. Run two-week tests instead of debates.

Write a brief hypothesis including a metric, target, owner, and rollback date. Change one variable only. Keep it if it moves the metric for two consecutive weeks without breaking a downstream metric. Otherwise, kill it fast and document the result.

7. Operate your board like a working session.

Run your board meeting like a working session, not a presentation. Send the board packet seventy-two hours in advance so members can prepare. Begin with a one-page summary that includes ARR, NRR, CAC payback, burn and runway, stage gate status, top risks, and top asks—give the board real substance. Structure the meeting flow as twenty minutes for highlights and misses, forty minutes for two deep dives, and twenty minutes for approvals and hiring. Close with a recap of decisions and owners, and circulate the meeting minutes within twenty-four hours.

8. Sending monthly investor updates on a cadence until Series B is a good default.

Send monthly investor updates on a consistent cadence until you reach Series B. Begin with a headline and stage status, followed by a metrics table showing ARR, NRR or GRR, CAC payback, win rate, sales cycle, and new ARR. Summarize what worked, what did not, and what you are changing next. Include key hiring and pipeline notes, along with your runway and funding plan. Close each update with three specific asks, each assigned to an owner.

9. Choose channel partners and intros with guardrails.

Choose channel partners and introductions carefully and apply clear guardrails. Do not build a channel program until you can reliably win direct deals. Pilot each partner relationship with a ninety-day mutual revenue target and a clear exit clause. Track sourced meetings, pipeline, wins, and sales cycle length by partner to measure real performance. Decline warm introductions that push you outside your ICP or distract from your current stage focus.

10. Watch for red flags.

Recognize red flags early to avoid wasted effort and misaligned relationships. Be cautious of advice that lacks a clear mechanism or a real example. Avoid partners or customers who push for heavy customization before proving first value. Decline introductions that generate work without pipeline. Watch for discounts or price changes that ignore established give–get rules. Steer clear of would-be advisors who seek a title instead of measurable outcomes.

Example

A Series A company with a $12K ACV took advice to "go enterprise" because a famous advisor could make intros. The pipeline looked big, and nothing closed. The team ran VRMON and found the mechanism was weak, and the sales cycle would eat their runway. We reset to midmarket, wrote a two-week message test, and returned a $24K target. Reply quality improved, the sales cycle shrank, and the company hit its numbers without hiring a field team.

Our Experience

We have followed shiny advice and paid for it. The turning point came when we applied the same filter to every outside idea that we use for our own ideas. By articulating and documenting the decision, setting RACI, and using two-week tests with rollback dates, we reduced our mistakes and debates. The board got calmer. The team moved faster.

Putting It All Together

Advice is input. Your system decides. Use stage gates from Chapter 2 to pick what to test now. Use ICP and SOM from Chapters 3 and 4 to ignore advice that drifts from your buyer. Use the process and RevOps backbone from Chapters 10 and 11 to run clean tests. Use the pricing rules from Chapter 9 to avoid discount theater. Use the reporting in Chapter 18 to show results without a narrative.

Companion Resources

WS27 supports chapter 14. It helps map investors, advisors, and decision rights for clean governance.

Download templates, worksheets, checklists, GTM-focused AI bots, and planning tools referenced in the book to help you apply the content directly to your go-to-market. The materials are updated periodically. Register for updates on our website's Resources page.

The Resources are available by using the QR code in the Companion Resources section at the beginning of the book.

15

BUILDING THE GTM ENGINE THAT SCALES

"What got you here won't get you there."
— *Marshall Goldsmith*

Insight

Scaling is not more of everything. It is more of the few things that already work, with guardrails that keep quality from collapsing under volume. If you scale before you demonstrate readiness, you multiply noise. If you scale after demonstrating readiness, you compound learning. Treat scale as an engineering problem: capacity, constraints, control loops, and transparent decision making.

Practical Walkthrough

Design a scale-ready GTM in nine steps.

1. Write a growth hypothesis.

Write a one-page hypothesis that states where incremental ARR will come from in the next 12 months: segments, offers, geos, channels. Add three falsifiable assumptions, and the metric each will move. Review the hypothesis quarterly. If reality disagrees, the hypothesis changes before the headcount does.

2. Build a capacity model.

Build a capacity model that starts with your target for new ARR this year and works backward to define the number of deals, opportunities, meetings, touches, and reps required. Use simple capacity guides: an SDR can handle sixty to one hundred quality contacts per day or ten to fifteen monthly meetings when list quality and messaging hold. An AE can manage twelve to eighteen active opportunities and twenty to thirty monthly meetings, depending on ACV and sales cycle. A Sales-Assist can support thirty to fifty active trials or onboarding checklists. Hire to relieve bottlenecks and add SDRs if meetings are the constraint, or AEs if opportunity load is the constraint, and Sales-Assist roles when activation stalls. Do not add managers until there are six to eight direct reports.

3. Shape the channel portfolio.

Shape your channel portfolio to stay balanced and scalable. No single source should account for more than sixty percent of meetings for two consecutive quarters. Add a new channel only after the primary one meets its targets for two complete sales cycles. For partners, run a ninety-day pilot with a mutual revenue target and an exit clause, and track meetings, pipeline, wins, and sales cycle length by partner.

4. Specialize the team deliberately with clear lanes.

Specialize the team deliberately and give each role clear lanes. The SDR is responsible for sourcing and qualifying. The AE owns discovery, proof of value, and the MAP. Specialists or CSMs join only for a defined proof of value or a Sales-Assist motion that accelerates the deal. Add Sales Enablement once you have five or more AEs, and bring in a Product Marketing manager when messaging changes often or new product features ship monthly or faster. Document decision rights in a short RACI covering pricing, discount bands, ICP edits, channel additions, and tooling.

5. Keep the tech stack small. Use a tech decision matrix. Before buying a tool, use a tech stack decision matrix. Don't add a new tool if you cannot quantify the benefit of adding it.

The tech stack decision matrix below helps you evaluate whether to add, replace, or delay a tool. It's designed to prevent stack sprawl by forcing clear decisions based on real needs.

Each column represents a type of tool (e.g., sequencing, data enrichment, or Customer Data Platform [CDP]). Each row walks through how to decide: what's the pain, are there non-tech fixes, who owns the decision, and when it will be reviewed.

The version shown here is an example. Your tech stack, constraints, and sales model will be different. Use this as an example to build your own. Fill in the matrix when a new tool is proposed or when revisiting existing ones. This process keeps your tech stack focused, accountable, and aligned with actual go-to-market needs.

Decide	Sequence: Add or Replace	Enrich Data	CDP or Warehouse
What are we trying to solve?	SDR output capped by manual sending	Low validity on new segments	5+ sources, weekly CSV merges
Non-tool Options	Tighten lists, rewrite openers, raise targets	Smaller ICP slice, manual research	Simplify metrics to what CRM can answer
Decision. Owner	Keep current tool. RevOps owns	Lightly enrich only after capture. SDR Lead owns it	Defer until readiness test reaches the pass threshold. CFO + RevOps own it
Review/ Sunset Date	Review in 60 days	Review in 30 days	Review in 90 days

Example of a Decision Matrix for Your Tech Stack.

6. Scale the operating cadence.

Scale the operating cadence to keep the team aligned and accountable. Hold a weekly pipeline review focused on stage accuracy, time in stage, stage-to-stage and overall conversion rates, win rate, median discount, sales cycle, ARR added, and one improvement with an assigned owner and rollback date. Set clear forecast rules that combine stage-based probability with MAP confidence, and report three views: commitment, best case, and total pipeline. Commitment requires a named buyer plan and confirmed dates. Review quality each week by sampling five call clips, five emails, and five MAPs.

7. Guardrails protect quality.

Guardrails protect quality and prevent the system from drifting. Enforce give–get pricing rules through a deal desk. Maintain SLA hit rates of at least ninety percent for inbound first touch and PQL to human, and keep stage accuracy above ninety-five percent. Monitor discount variance weekly, and if the median discount increases, pause tests and retrain the team. No deal should advance without a defined next step and date.

8. Plan for land-and-expand.

Plan for land-and-expand growth by defining three clear expansion triggers that signal revenue potential: seat growth, usage thresholds, and new teams being invited. Assign ownership of expansion plays to Sales-Assist and Customer Success, each operating on a structured 30–60–90-day plan. When measurable outcomes appear, ask for referrals and case studies at days thirty and ninety to capture momentum and social proof.

9. Future-proof for scale.

Future-proof the business so it can scale without breaking. Default to reversible changes and test them with small cohorts before rolling out broadly. Ensure observability by requiring every automation to write to a log with a named owner. Maintain strong data governance through a quarterly review of the glossary, pick lists, and access permissions. Apply AI hygiene using AI only to summarize, draft, or detect anomalies—never to invent facts. Store all prompts and outputs for audit. Manage vendor risk with a one-page exit plan for every critical tool.

Example

A team jumped from 5 sellers to 18 in a quarter. Meetings increased, win rate fell from 26% to 13%, median discount rose to 21%, and the sales cycle lengthened by 20 days. The team paused hiring, wrote a growth thesis, and rebuilt the capacity model. We limited each AE to 15 active opportunities, recentered on one ICP and one offer, and enforced MAP on day one of discovery. We eliminated a partner program that generated meetings but yielded no wins. In eight weeks, stage accuracy hit 96%, discount fell below 12%, sales cycle shortened by 14 days, and forecast error dropped to under 10%. Only then did we add headcount.

Our Experience

We have watched tools become increasingly complex, and managers add meetings that do not move metrics. The moves that unlock scale are not tech platforms. They agree to weekly experiments, measure them, and keep the targets high. The companies that win at scale make small decisions quickly and big decisions slowly.

Putting It All Together

Everything you built so far becomes a flywheel: stage gates from Chapter 2, SOM and ICP from Chapters 3 and 4, messaging and channels from Chapters 5 through 7, the first sales process from Chapter 10, and the RevOps backbone from Chapter 11. Pricing and give-get rules from Chapter 9 protect the margin. Hiring from Chapter 12 adds capacity without breaking quality. Next, we move to how the founder's role changes once the engine runs without you.

Companion Resources

WS28 and WS33 support Chapter 15. They guide GTM road mapping and the architecture decision grid.

Download templates, worksheets, checklists, GTM-focused AI bots, and planning tools referenced in the book to help

you apply the content directly to your go-to-market. The materials are updated periodically. Register for updates on our website's Resources page.

The Resources are available by using the QR code in the Companion Resources section at the beginning of the book.

SECTION IV

REAL-WORLD EXECUTION

16

THE FOUNDER'S
ROLE AFTER GTM FIT

"Work on the business, not in it."
— *Michael E. Gerber*

Insight

Once GTM fit is confirmed, your job changes. You stop being the best seller and become the system designer. If you keep closing the biggest deals, approving every discount, and rewriting every email, you will cap the company's growth at your personal capacity. The role now is to set ambition and sequence, protect focus, hire and develop leaders, enforce decision rights, and run a cadence that turns learning into action without you in every room.

Practical Walkthrough

Redesign your week and your personal operating system so the company scales beyond you. For example:

1. Redesign your calendar.

Redesign your calendar to reflect your highest priorities. Spend 40% of your time recruiting and one-on-one meetings with functional leads, 30% with customers and market discovery focused on your highest-signal ICP, 20% on product and strategy synthesis with the executive team, and 10% on capital, board, and key external partners. Block time each week for founder-only thinking, and attend status meetings only when your presence is required.

2. Put in place a simple management cadence.

Establish a simple management cadence that drives decisions rather than updates. Hold a weekly review led by operations to check stage accuracy, time in stage, conversion waterfall, win rate, median discount, sales cycle, ARR added, and one improvement with a named owner and rollback date. Run a product–GTM sync to discuss what buyers said, what will be tested, and what will be shipped. Review hiring progress by tracking open roles, candidate pipeline, scorecards, and offers. End with a change review to approve or retire any tests that affect pricing, routing, or data definitions. Attend these meetings only when a decision is required, not to narrate.

3. Set decision rights and escalation rules.

Set clear decision rights and escalation rules to speed up execution. Use a short RACI to define ownership for pricing, discount bands, ICP edits, channel additions, and tech stack changes. Classify every choice as either two-way (reversible) or one-way (difficult to undo). Set SLAs for decisions: make two-way calls within forty-eight hours and require a short brief and a date for one-way calls. Once a decision is made, disagree and commit (no passive-aggressive vetoes).

4. Make the documented changes real.

Turn documented changes into real operational improvements. Write a short brief for every material change, including the hypothesis, metric, target, owner, start date, and rollback date. Limit each team to one change per week. Keep a change only if the metric improves for two consecutive test cycles without negatively affecting any downstream metric. Publish every change so the team can see what was adjusted and why.

5. Raise the target for managers.

Raise the performance bar for managers as the organization grows. Player–coach managers should oversee no more than six to eight direct reports. Each manager is responsible for running the cadence, coaching through call and email reviews, and maintaining data hygiene. Promote only when the process operates smoothly without their direct involvement, and never give out titles as a retention tactic.

6. Hire leaders at the right time.

Hire leaders only when the time and conditions are right. Do not hire a VP to find product–market fit; hire one to scale a motion that already works. Create scorecards for leaders that measure their ability to hire people who hit targets, install effective systems, move key metrics, and make necessary replacements. Hold skip-level meetings monthly to detect process drift and uncover quiet misalignment before it spreads.

7. Reduce founder dependency.

Reduce founder dependency to build a scalable sales motion. Track the percentage of wins involving you and aim to keep it below twenty percent. Measure how often you join late-stage calls and reduce that rate through strong MAP discipline and clear proof of value. Each quarter, create a "stop list" of tasks you will no longer handle and assign new owners for each.

8. Be the storyteller, not the hero.

Write a one-page strategy memo each month outlining where growth will come from, what was learned, what will change, and what will stop. Lead with numbers and follow with narrative context. In board meetings, focus on working sessions rather than presentations, and use the time to ask for help on one or two specific decisions.

9. Protect focus.

Protect focus by limiting priorities and aligning work to measurable impact. Choose only three key metrics to move each quarter and treat everything else as backlog. Eliminate side projects and decline introductions or partnerships that do not support your ICP or current stage plan. Keep the roadmap anchored to tagged customer signals rather than internal opinions.

10. Future-proof the backbone.

Future-proof the company's operational backbone to keep it reliable as you scale. Review definitions, discount bands, and routing rules quarterly. Apply light governance for data and AI use by storing prompts and outputs for audit, prohibiting fabrication, and assigning ownership for any model that affects routing or scoring. Maintain a one-page vendor replacement plan for every critical tool.

Example

A founder at a $4M ARR business still closed every large deal. Discounts were creeping up, sales cycle length grew, and AEs waited for approvals. We moved approvals to a deal desk with give-get rules, and set a target that fewer than 20% of wins would involve the founder. The founder shifted time to hiring managers and running a weekly change review. The median discount fell 8% in two quarters, stage accuracy hit 96%, and forecast error dropped to under 10%. The biggest deals still close, but now without the bottleneck.

Our Experience

The hardest change is letting go of work you are great at. When we forced founders to publish a stop list, hand off approvals, and show up only when a one-way decision was needed, teams grew faster and leaders grew up. The culture shifts from heroic saves to quiet execution.

Putting It All Together

This chapter sits on top of the whole book. Stage gates from Chapter 2 tell you what to prove next. SOM and ICP from Chapters 3 and 4 keep focus. Messaging and channels from Chapters 5 through 7 fill the top of the funnel. The sales process and RevOps backbone from Chapters 10 and 11 make execution teachable. Pricing from Chapter 9 protects the margin. Chapter 15 gave you the scale plan. Next, Chapter 17 will show how to keep the system coherent as it grows.

Companion Resources

WS34 and WS22 support Chapter 16. They help the founder shift roles, delegate, and reset the operating cadence.

Download templates, worksheets, checklists, GTM-focused AI bots, and planning tools referenced in the book to help you apply the content directly to your go-to-market. The materials are updated periodically. Register for updates on our website's Resources page.

The Resources are available by using the QR code in the Companion Resources section at the beginning of the book.

17

THE SYSTEM IS THE STRATEGY

"The essence of strategy is choosing what not to do."
— Michael Porter

Insight

Winning companies do not chase every idea. They run a simple operating system that selects what to do now, ignores the rest, and compounds learning. Your strategy is the set of rules you follow every week: one ICP, one offer, one teachable process, one source of truth, clear targets, and a change log you actually use. When the system is healthy, growth is a by-product.

Practical Walkthrough

Build a coherent GTM system in 30 days. Keep or kill based on targets.

Stage Gate Audit

Use Chapter 2. Mark each stage gate green, yellow, or red. Document the missing targets and the change that would let you hit the target.

Commit to One ICP and One Offer

Use Chapters 3 and 4 to pick a single segment for this cycle. Define triggers, access, and economics. Lock the scope for 30 days.

Freeze Messaging and Proof of Value

Select the winning message variant from Chapters 5 and 10 and one proof of value. Create and save one-pagers, the MAP, and recap templates where AEs can find them quickly.

Pick Exactly Two Channels

From Chapters 6 and 7, choose the two highest-yield channels for this ICP. Hold all others. Set two weeks to achieve targets and a rollback date.

Pricing Guardrails

From Chapter 9, publish the list, discount bands, and give-get rules. Route the exceptions to a named approver. Track the median discount weekly.

Tighten the RevOps Backbone

From Chapter 11, enforce SLAs, required fields, next steps with dates, and stage exit criteria. Build the one-page CEO dashboard.

Run the Cadence

Run a weekly pipeline review with numbers from the CRM: one change per team per week. Every change has an owner, metric, targets, and a rollback date.

Align Finance and Hiring

Use the SOM and SQL models from Chapter 3. Lock a 13-week hiring and spend plan. Add a simple risk register with owners.

Set Success Targets

Stage accuracy ≥95%.
Median discount at or below guardrail.
Outbound or inbound targets hit two weeks in a row.
Sales Cycle time is stable or improving.
The CEO dashboard refreshes in under 60 minutes from the CRM.

Weekly One-Page System Scorecard

Green means goals are held two weeks in a row. Yellow means watch. Red means stop volume and fix.

Targets

- ICP focus: one ICP and trigger locked.
- Messaging: winner variant documented.
- Channels: two active, targets defined.
- Process: MAP used on all late stages.
- RevOps: 100% next-step hygiene.
- Pricing: median discount.
- SLA: inbound 5 min, PQL 2 hours.
- Reporting: CEO dash updated in under 60 min.

Assign an owner to each target. Score the results, and define one change you'll make for next week.

GTM Maturity Model

Use this to align the executive leadership team and the board *(see Chapter 2 for complete GTM Maturity Stage definitions)*.

ZERO TO SOMETHING

At this stage, the focus is on proving that genuine buyers will pay and that at least one repeatable path to a win exists. Sales are founder-led, channels are opportunistic, and only a basic CRM is needed. Don't follow a temptation to avoid a CRM at this stage. Valuable data needs to be collected and put into a CRM. Usually, a free tier of a top CRM will get the job done.

Your ICP is still a hypothesis. You're hunting for early behavioral signals that indicate intent. Messaging is direct and personal, often written by the founder and refined through conversations. Channels are experimental and are used mainly to test reach and message fit. The sales process is simple: one clear path from first contact to closed-won, designed primarily to learn what breaks. RevOps is light-touch, focused on data hygiene and consistent definitions within a basic CRM.

Pricing is often custom and flexible, since each deal helps you understand perceived value. Reporting happens in spreadsheets rather than dashboards, and AI or data tools are optional (helpful for drafting but not yet essential). The goal is learning fast, not scaling.

SELLING BEFORE SCALING

Now, the goal is repeatability to show that the motion works every time and handoffs hold steady. You're transitioning from founder-driven sales to a team that can operate the playbook independently.

The ICP evolves into a behavioral profile with clear triggers that signal when to engage. Messaging is still tested but structured; by now, you've tried at least three variants and know which performs best. Channels narrow to two or three that reliably deliver the pipeline. The sales process becomes explicit: one defined path with exit criteria, a mutual action plan, and measurable conversion points.

RevOps introduces service-level agreements, shared definitions, and regular hygiene checks. Pricing adopts basic lists and give-get rules to manage discounts. A hiring scorecard emerges to support new AEs, and reporting shifts into the CRM with a living CEO page. AI may assist in summaries and follow-ups, but a data warehouse isn't mandatory. Consider it only if the data volume passes the threshold where your data volume, complexity, or business risk exceeds what your CRM or spreadsheets can reliably handle.

HIRING WITHOUT BREAKING

Here, the aim is consistency under load. You're no longer proving the motion. You're training others to run it without regression.

ICPs are now divided into two or three validated segments, each with buying rules. Messaging moves to product marketing for cadence, proof points, and a growing library of value stories. Channels are limited to a few proven ones with clear ROI caps. The sales process matures into team-wide enablement, with trained reps forecasting accurately and following documented playbooks.

RevOps takes ownership of routing, automation, and deal-desk efficiency. Pricing is tested methodically by segment. Reporting becomes scorecard-based, giving every team visibility into performance. AI models are explainable and targeted to pipeline health, and a minimal data stack is in place to support scale.

SCALING WITH CONFIDENCE

At this stage, the business is focused on efficiency at volume, tightening margin, payback, and governance while preserving agility.

The ICP is reviewed quarterly and tracked as part of a managed portfolio. Messaging evolves continuously through small, measured lifts. Channels are diversified to prevent dependency; no single one should drive more than 60 percent of sales meetings. The sales process includes guardrails for time-in-stage, regular audits, and ongoing enablement.

RevOps now owns the whole data warehouse and governance cycle, ensuring lineage, integrity, and visibility across teams.

Pricing and margin are tuned for the defined CAC payback period. Hiring ratios remain balanced with healthy manager-to-IC loads. AI and data operations are fully auditable and reviewed quarterly, reporting shifts to automated anomaly alerts and weekly insight reviews rather than manual merges.

USING THE MODEL

Revisit this framework once per quarter with your leadership team. Ask: *Which stage are we truly in?* Then align hiring, pricing, and RevOps maturity to that stage before pursuing further scale.

90-Day Operating Plan Template

Plan by month. Keep it on one page.

Month 1

- Focus on ICP and offer:
- Channel experiments:
- System fixes :
- Hiring decisions:
- Targets to hit:

Month 2

- Keep or kill from month 1:
- New test:
- Enablement update:
- Pricing tune-up if needed:

- Targets to hit:

Month 3

- Scale decision: headcount or spend:
- Second ICP consideration: yes or no:
- Board-ready results and next thesis:
- Targets to hit:

Risk Register (rolling)

- Risk, owner, mitigation, trigger to act.

Our Experience

Teams that were "busy" turned into teams that learned when they limited changes, enforced exit criteria, and ran one cadence with a public change log. In one example, the team paused three channels, locked one ICP and one offer, enforced MAP, and routed PQLs within two hours. In six weeks, stage accuracy hit 96%, median discount fell 7%, the sales cycle shortened by 12 days, and forecast error dropped to under 10%. Nothing fancy, just a system that people could run.

Putting It All Together

This chapter is the glue. It connects ICP, message, channels, process, RevOps, pricing, hiring, and reporting into one weekly loop. When something slips, the system tells you what to fix next. When it holds, you scale the things that work. Next, Chapter 18 turns this into reporting that leaders and boards can trust.

Companion Resources

WS01, WS29, and WS36 support Chapter 17 and drive the system audit, stage-gate review, and maturity check.

Download templates, worksheets, checklists, GTM-focused AI bots, and planning tools referenced in the book to help you apply the content directly to your go-to-market. The materials are updated periodically. Register for updates on our website's Resources page.

The Resources are available by using the QR code in the Companion Resources section at the beginning of the book.

18

REPORTING THAT BUILDS CONFIDENCE

"When you can measure what you are speaking about, and expressit in numbers, you know something about it."
— *Lord Kelvin*

Insight

Reporting is how you run the company; it's not a slide deck. Good reporting makes the next decision obvious and fast. Bad reporting creates arguments and delays. Build a small set of definitions everyone trusts, and a cadence that turns numbers into action.

Practical Walkthrough

1. Create a reporting system that leaders and boards trust.

Determine what decisions you need to make:

- Weekly: Keep or kill test, where to add effort, what to unblock.
- Monthly: hiring and spending, channel scale up or down, pricing adjustments.
- Quarterly: add a segment, open a geo, raise or conserve capital.
- Map each decision to one metric. If a metric does not support a decision, drop it.

2. Publish a metrics usage and ownership document. Create a one-page table with definition, formula, source, owner, and refresh cadence.

ARR is the sum of the active ARR book of business (contracted annual recurring revenue). ARR is based on the CRM and invoices issued. Finance owns the number, and it should be refreshed weekly.

Net New ARR is (new – churn + expansion). It is based on a given period's bookings net of churn. Net new ARR is based on the CRM and invoices issued. Finance owns the number, and it should be refreshed weekly.

Win Rate is closed won deals ÷ opportunities (W ÷ Opps). The CRM is the SSOT. RevOPs owns this number, and it should be refreshed weekly.

Median Sales Cycle is the days from SQL to close won. If you have segments with significantly different sales cycles, break them out into cohorts, or your sales cycle won't match the reality of either segment, especially if you use mean instead of median. The SSOT is the CRM. RevOps owns the sales cycle measurement, which should be updated weekly.

Median Discount is the percentage of below-target-price at close. Like the sales cycle, measure by segment. The SSOT is the CRM. The deal desk (or sales if no deal desk) owns the measurement, which should be updated weekly.

SQLs per month is the count of new leads accepted by sales (SQL). The SSOT is the CRM; RevOps owns the number, which should be updated weekly.

SLA Hit Rate is the rate at which SLAs are met. It is calculated as hits ÷ total handoffs. The SSOT is the CRM. RevOps owns the calculation, which should be updated weekly.

Stage Accuracy is a metric measuring whether deals are in the correct stage. Usually, this must be audited manually. If you have many deals, then take representative samples. The SSOT is the CRM. Sales or RevOps owns it, and it should be updated weekly.

Lead Source Mix is the ratio of where leads come from: outbound, inbound, PLG, or partner. Define the percentage per source. The SSOT is the CRM. RevOps owns it, and it should be updated weekly.

Activation and Trial to Paid is related to PLG conversions. This is event-based and requires product analytics. Usually, Sales or Customer Success owns this number, which should be updated weekly.

NRR (or GRR) measures net or gross revenue retention. Invoicing is the SSOT; the calculation is owned by finance. NRR and GRR should be updated monthly.

$$GRR = (MRR_0 - \textit{Churned MRR} - \textit{Contraction MRR}) \div MRR_0$$

$$NRR = (MRR_0 - \textit{Churned MRR} - \textit{Contraction MRR} + \textit{Expansion MRR}) \div MRR_0$$

CAC payback is the months to recover CAC. The formula is CAC ÷ Gross Margin per month. Finance usually owns this number, and it should be updated monthly.

3. Build a one-page CEO dashboard—three blocks on one slide.

Build a one-page CEO dashboard that fits on a single slide and shows performance, execution, and quality at a glance. In the performance block, track ARR, net new ARR, NRR or GRR, and CAC payback. In the execution block, include SQLs, win rate, median sales cycle, median discount, and source mix. In the quality and risk block, show stage accuracy, SLA hit rates, forecast accuracy, and the top three loss reasons. Pull all data directly from the CRM or finance systems—no manual updates. Automate refreshes, and if it takes more than sixty minutes to update, fix your data capture or definitions.

4. Give every team a scorecard with one screen per team. Red, yellow, or green against agreed targets.

Give every team a scorecard displayed on a single screen, showing red, yellow, or green status against agreed targets. For SDRs, track valid rate, reply rate, positive reply rate, meeting rate, and conversion from meeting to opportunity. For AEs, measure discovery quality, MAP usage, stage exit accuracy, win rate, sales cycle, and discount levels. For Customer Success, track time to first value, expansion at days thirty and ninety, NPS by cohort, and active risk flags. For PLG, monitor activation events, trial-to-paid conversion, PQL volume, and Sales-Assist response time.

5. Run the forecast like an operating system.

Run the forecast like an operating system that drives accountability and precision. Use three lanes: pipeline, best case, and commit. A commit deal must have a named buyer plan and confirmed dates within a MAP. Roll up forecasts by stage and by cohort, keeping SMB and enterprise separate to avoid distortion. Measure forecast accuracy at thirty, sixty, and ninety days, and focus on improving the underlying system rather than refining the narrative.

> The numbers of your business must support the story you tell about your business.
>
> If the numbers and the story don't match, either you're telling the wrong story or the business isn't performing as expected.
>
> Never fake the numbers to match the story. Hit your numbers or change your story.

6. Cadence forces decisions.

Use cadence to force real decisions and continuous improvement. In the weekly pipeline review, start with the CEO dashboard, choose one change with an assigned owner and rollback date, and keep the meeting to forty-five minutes. Avoid retelling the past. In the monthly review, evaluate capacity and spend versus plan, the channel portfolio, hiring progress, and pricing guardrails. In the quarterly review, assess stage gate status from Chapter 2, define the next quarter's hypothesis, and identify key risks with mitigation plans.

7. Make it explainable.

Make your system explainable so every number can be trusted. Store all definitions in a shared glossary, and assign a clear formula and owner for each metric. Require every automation to write to a log so that any change in results can be traced. Use cohorts and medians to reveal real trends instead of relying on averages that can hide problems. If you plan to add a data warehouse, complete the readiness test from Chapter 11 first and begin with a small, focused scope.

CFO Guardrails

Set clear CFO guardrails to maintain financial integrity and consistency. Keep bookings and ARR separate. Never combine them in the same chart; side-by-side in different charts is fine. Calculate CAC payback using actual gross margin and include the effect of discounts. Track discount drift, and if the median discount increases, pause experiments and retrain the team. Build board packs using the same definitions and data sources used in weekly reviews to ensure full alignment.

Example

A Series A company spent six hours a week merging spreadsheets. Marketing showed leads, Sales showed pipeline, and Finance showed GAAP revenue. None matched. We guided the team in publishing a metric canon, built a one-page CEO dashboard from CRM and billing, and moved the forecast to commit rules tied to MAPs. In four weeks, refresh time dropped to 20 minutes, forecast error fell to under 10%, and board meetings became working sessions.

We have seen CEOs fired because different teams reported different metrics or defined metrics differently, even when using the same naming conventions. The CEOs weren't fired because teams used different metrics. The CEOs were fired because the board and investors kept catching inconsistencies that the CEO couldn't explain in real time. It often took weeks for the CEO to find the actual numbers because none of the teams spoke the same metrics language.

Our Experience

When reporting turns into unaligned storytelling, trust dies. The team moved faster when we cut metrics to the few that drive decisions, published the formulas, and enforced the cadence. The biggest unlock was a shared language and a rule that every meeting ends with an aligned understanding of the numbers.

Putting It All Together

This chapter makes the whole system visible. It sits on the RevOps backbone from Chapter 11, reflects pricing guardrails from Chapter 9, and measures the quality targets you set in Chapters 5 through 8 and 10 through 12. It feeds Chapter 19 when you add AI to accelerate good judgment. If the dashboard does not change your actions this week, that is not reporting. It is a decoration.

Companion Resources

WS30 and WS29 support Chapter 18. They define the CEO dashboard, reporting cadence, and KPI dictionary.

Download templates, worksheets, checklists, GTM-focused AI bots, and planning tools referenced in the book to help you apply the content directly to your go-to-market. The materials are updated periodically. Register for updates on our website's Resources page.

The Resources are available by using the QR code in the Companion Resources section at the beginning of the book.

19

OPERATING IN AN AI-ACCELERATED GTM

"The future is already here. It's just not evenly distributed."
— *William Gibson*

Insight

AI will not fix a broken GTM, but it will amplify whatever you already run, for better or worse. Use it to shorten sales cycles, raise quality, and focus humans on decisions. Keep rules simple: assist before automating, explain before trusting, log everything, and never let a model fabricate facts to a customer. Treat AI like code with an audit trail and a rollback plan.

Practical Walkthrough

Build an AI operating system in 30 days.

ASSIGN THE RIGHT LEVEL OF AUTONOMY

Assist: Use AI to draft, summarize, or suggest next steps. These are low-risk use cases that can be deployed quickly.

Suggest: Let AI score, rank, or recommend options, but keep a human in the loop for final approval. Set clear threshold bands for when human review is required.

Decide: Allow AI to take automated actions only after the Assist and Suggest tiers have proven reliable and delivered measurable value.

EVALUATE HIGH-LEVERAGE USE CASES

Call and demo summaries tied to stage and MAP updates.

Discovery notes turned into buyer language follow-ups.

Outbound research briefs that pull firmographics and public triggers.

Variant generation for messaging tests with human editing.

Anomaly detection on SLAs, stage creep, discount drift.

PQL hints from activation events with suggested next touch.

Pipeline explainers that show which deals are at risk and why.

Objection analysis with links to proof of value assets to use next.

Security questionnaire is prefilled using your published answers.

Forecast explanations that highlight pattern shifts by segment.

WRITE A ONE-PAGE AI POLICY

Truth: no fabrication. Use only verified data and cite the source in internal outputs.

Privacy: no personally identifiable information to external models unless approved.

Human-in-the-loop: customer-facing content is consistently reviewed by a person.

Logging: store prompts, outputs, and approvers.

Ownership: name owners for prompts, models, and access.

Security: vendor access is the lowest viable privilege and is time-bound.

CREATE A SMALL PROMPT LIBRARY

Create one folder per function (e.g., SDR, AE, PMM, CSM, RevOps, Finance). Each prompt should have a goal, inputs, constraints, example output, and a reviewer. Keep prompts short with examples that reflect your ICP and tone. Version the prompts like code. Retire the ones that do not move a metric.

VALIDATE THE QUALITY

Pick 100 records per use case. Define a scorecard: accuracy, relevance, tone, and actionability on a 1 to 5 scale. Keep only if the median score is at or above 4 for two weeks and a downstream metric improves. Add a red team pass: try to break it with edge cases and outdated info.

WORK WITH REVOPS, NOT AROUND IT

Connect AI tools to your CRM. This can be native or an integration, but don't add complexity and manual steps.

Whenever a meeting or sales call summary is logged (whether by AI or manually), it should automatically generate follow-up actions.

When the system flags a PQL, it should automatically create an assigned task in the CRM for follow-up.

When something breaks pattern (e.g., a deal stalls past SLA, a response time is overdue, or a conversion rate dips), the alert should include both the data record and the service-level agreement (SLA) it violated.

SET GUARDRAILS FOR ROUTING AND SCORING

- Use the Suggest tier first for lead scoring.
- Require an explanation field in plain language.
- Humans can override with a reason code.
- Review overrides weekly. If humans override more than 30%, fix or retire the model.

VENDOR AND ARCHITECTURE CHOICES

If the Chapter 11 warehouse test fails, defer centralization and keep data inside your CRM and help desk. When you are ready, the minimal stack is a data warehouse, ELT, a modeling layer, and reverse ETL.

Use a vendor checklist. Create a standardized due diligence form or spreadsheet to evaluate every vendor's security, compliance, and exit readiness.

GOVERN CHANGES

Govern AI changes with the same disciplined cadence used for all operational updates. Each AI change must include a clear hypothesis, metric, target, owner, start date, and rollback date. Limit teams to one AI change per week, and keep a change only if it improves a real metric without negatively affecting any downstream results.

TRAIN THE TEAM

- 60-minute workshop per function with three live prompts, two bad outputs, and the review rules.
- Set a style guide for tone and claims.
- Ensure your team is only paying vendors for legitimate, policy-compliant data or AI usage.

KEEP OR KILL RULES

- Keep tech tools that save time and improve quality reproducibly.
- Keep tools after testing that you don't regularly need to bypass, since they improve the target metric.
- Do not commit to the new tool without two successful cycles and a clear rollback plan.

Example

A team used AI to summarize calls and draft follow-ups. We integrated the summaries into the CRM with next steps and MAP updates, and blocked auto-send. SDRs used research briefs and message variants but still edited for tone and facts. Over six weeks, follow-up time fell from 26 to 4 hours, MAP usage rose 20%, and stage accuracy hit 96%. When we tried to use automated meeting booking from chat, no-show rates spiked. We rolled back in a day and added a human confirmation step.

Our Experience

AI pays when it removes toil and raises quality at the edges of the funnel. It fails when teams try to skip the backbone or automate their judgment. The best wins came from short prompts tied to a clear outcome, a reviewer, and a metric. The most significant change was anomaly detection in RevOps: models flagged stale next steps and stage creep before forecast meetings, and managers coached sooner.

Putting It All Together

AI rides on the system you built in earlier chapters. ICP and messaging feed prompts. Both outbound and inbound benefit from variant generation and research assistance. The sales process and MAP give structure to summaries and next steps. RevOps owns truth, logging, and routing. Reporting shows where AI helps and where it hurts. The maturity model in Chapter 17 tells you when to push further and when to pause.

Companion Resources

WS31, WS32, WS33 support chapter 19. They cover AI readiness, use-case backlog, and policy governance.

Download templates, worksheets, checklists, GTM-focused AI bots, and planning tools referenced in the book to help

you apply the content directly to your go-to-market. The materials are updated periodically. Register for updates on our website's Resources page.

The Resources are available by using the QR code in the Companion Resources section at the beginning of the book.

FINAL THOUGHTS

Run the system you built. Revisit the stage rubric each quarter. Keep the scorecard honest. Add volume only after repeatability shows up in other people's hands. If revenue rises when you leave the room, you did it right.

You Will Never Feel Ready

You'll be tempted to think that once the ICP is clearer, once the messaging is tighter, once you raise the next round, then you'll finally commit to building the system. But don't wait. The companies that win do not have the cleanest decks or flashiest tactics. They're the ones who start, adjust, and keep going while everyone else is still strategizing. GTM is not a one-time project. It's a habit. You will be told to do more. More channels. More personas. More verticals. More tools. More features. More noise. Say no.

Start-ups die of indigestion, not starvation. Your job is to protect the signal, not chase every shiny object that someone with a title tells you "could work." More is not better. Better is better.

You Will Want to Step Away

As traction builds, you'll feel pulled into ops, investors, HR, finance, and culture. You'll feel like stepping back from GTM is a reward for early success. Please don't do it. Stay close to the signal. You don't have to run every call, but you must know how the machine runs and where it breaks. That's the difference between a company that coasts and one that compounds.

You Built Something Real

Most founders never get this far. They ship code, raise money, and build a team, but never build the system. You did. That means you're no longer guessing. You're learning. You're improving. You're not just playing the start-up game. You're playing to win. And now, you've got the field manual to keep going.

One last reminder:

> *"The way to get started is to quit talking and begin doing."*
> — *Walt Disney*

We're with you. Let's go.

APPENDICES

APPENDIX A

THE GTM CANON

This is the starting list of core GTM metrics, practices, and stage gates. Every company adapts, but use these as your minimum operating backbone to begin with, and over time, you will develop your own canon.

ICP Clarity

- **Definition:** Your ideal customer profile is written down, shared, and agreed upon across Sales, Marketing, and Product.
- **Details:** Must include firmographics (industry, size, geography), technographics (tools in use), roles/personas, and specific pain points you solve.
- **Target:** At least 80% of the pipeline aligns with ICP.
- **Cadence:** Review and update quarterly.

Message Testing

- **Definition:** Outbound and inbound messages are tested for response and conversion.
- **Details:** Track reply rate on sequences, open rate on email subject lines, and positive vs. negative replies.
- **Calculation:** Reply rate = positive replies ÷ total delivered messages.
- **Target:** ≥10% positive reply rate on outbound sequences.
- **Cadence:** Review weekly; update sequences monthly.

Pipeline Hygiene

- **Definition:** Active pipeline free of stale or unqualified deals.
- **Rules:**
  - No deal is idle for more than 14 days without activity.
  - All deals must have clear next steps logged.
  - Close dates updated weekly.
  - Prioritize winnable high-value deals.
- **Measurement:** Percentages of deals compliant with hygiene rules.
- **Target:** ≥90% compliance.
- **Cadence:** Weekly review.

Stage Gates

- **Definition:** Documented entry/exit criteria for each pipeline stage.
- **Examples**
  - **Qualification:** ICP fit + discovery call complete.
  - **Proposal:** Solution validated + budget confirmed.
  - **Commit:** Stakeholders aligned + contract in draft.
- **Measurement:** Percentage of deals meeting criteria before stage advancement.
- **Target:** 100% adherence (no "happy ears" advancing).
- **Cadence:** Validate at weekly pipeline reviews.

Handoff SLAs

- **Definition:** Clear agreements for passing opportunities across teams (SDR → AE → CS).
- **Details:**
  - SDRs must log all discovery notes in CRM within 24 hours.
  - AEs must accept/reject within 48 hours.
  - CSM must receive the kickoff doc before a customer call.
- **Measurement:** SLA compliance rate = # compliant handoffs ÷ total handoffs.
- **Target:** ≥95% compliance.
- **Cadence:** Monthly review.

Forecast Discipline

- **Definition:** Forecasts tied to data, reviewed consistently, and tracked against outcomes.
- **Details:** Weekly forecast call with pipeline owners; variance analysis of commit vs. actual.
- **Calculation:** Forecast accuracy = $1 - |(\text{commit} - \text{actual})/\text{actual}|$.
- **Target:** ≤20% variance.
- **Cadence:** Weekly forecast review, monthly accuracy report.

Price Realization

- **Definition:** Ability to capture intended value in deals.
- **Details:** Compare contracted ARR to list price or target pricing.
- **Calculation:** Price realization = actual ARR ÷ list ARR.
- **Target:** ≥90%.
- **Cadence:** Quarterly review with finance.

Retention

- **Definition:** Track both logo retention and NRR.
- **Formulas:**
- Logo retention = (logos retained ÷ logos at start of period).

- NRR = (Starting ARR + expansion – churn – contraction) ÷ (Starting ARR).
- **Targets:** Logo ≥90%; NRR ≥100%.
- **Cadence:** Monthly and quarterly reviews.

APPENDIX B

PROCESS AND TOOLS BY GTM MATURITY

Stage 1: Zero to Something

PROCESS FOCUS

Prove someone will pay. Document an ICP hypothesis. Capture first 3–5 wins. Founder-led messaging. Ad hoc process.

CORE TOOLS

CRM (basic: accounts, contacts, opportunities, activities)
Calendar + video + recording
Shared docs for ICP notes, MAPs, and exit criteria

Stage 2: Selling Before Scaling

PROCESS FOCUS

Motion repeats. ICP sharpened. Stage gates + MAPs in place. SLAs enforced. Handoffs hold.

CORE TOOLS
CRM with hygiene rules and required fields
Sales engagement platform (sequencing, call/email logging)
Light data enrichment (basic firmographics)
Contract / e-sign tool

Stage 3: Hiring Without Breaking

PROCESS FOCUS
Non-founder AEs win. Forecast accuracy in range. Team trained. Playbook in use. Enablement emerges.

CORE TOOLS
CRM with enforced stage gates and loss reasons
SEP + enrichment + routing automation
Enablement repository (call library, objection bank, proof of value library)
Forecasting/reporting add-on or business intel layer
Billing + contract integration

Stage 4: Scaling With Confidence

PROCESS FOCUS
Efficiency holds at volume. CAC payback tracked. Churn/NRR stable. Portfolio diversification. Governance live.

CORE TOOLS

CRM as source of truth (integrated with contract + billing)

SEP, enrichment, and enablement are fully operational

Data warehouse (if readiness test reaches the pass threshold)

ELT + reverse ETL for reporting

Deal desk automation + approval workflow

Governance tools (data lineage, access logs)

APPENDIX C

B2B SALES MODEL COMPARISON

Sales Model: Founder-Led

Best for: Early-stage, pre-GTM fit
Pros; High signal, flexible, fast learning
Cons; Not scalable, emotionally taxing
Notes; Essential for learning message, ICP, and sales motion

Sales Model: SDR + AE

Best for: Mid-stage, higher ACV, outbound-driven
Pros; Clear roles, scalable process
Cons; Requires a tight process and enablement
Notes; Works best after messaging and ICP are validated

Sales Model: PLG

Best for: Self-serve tools, low-friction onboarding
Pros; Efficient, viral loops are possible
Cons; Requires product depth and usage triggers
Notes; Still benefits from outbound and CSM
support for conversion and upsell

Sales Model: Inbound/ Marketing

Best for: Markets where buyers search and educate
online
Pros; Scalable, content-efficient
Cons; Long ramp time, less control early
Notes; Layer in after outbound traction, with a
proven message

Channel/ Partner Sales

Best for: Niche or global expansion
Pros; Access to trusted networks, lower CAC
Cons; Less control, slower feedback loops
Notes; Only works with strong ICP clarity and
partner alignment

Enterprise/ Field Sales

Best for: Complex, high-ticket solutions
Pros; Large deal sizes, deep relationships

Cons; Long cycles, high CAC, significant support needed

Notes; Requires a strong GTM fit; use only when sales maturity is high

APPENDIX D

VISUALIZING METRICS

Metrics driven by data and key performance indicators (KPIs) are integral for tracking progress, spotting trends, and making educated decisions. Yet, many CEOs regard metrics and KPI tracking as a pure admin chore, only for reporting purposes. This view ignores the crucial role metrics play in helping CEOs and managers run their companies efficiently. Most start-ups that are rigorous and metrics-focused grow faster and achieve more success.

Effectively showcasing your data through charts, graphs, and tables is just as crucial as data collection. You need a clear and comprehensible view of the figures driving your business to draw insights from your data and articulate your company to others.

The following charts and tables are practical tools to convey what's happening in your business. Your approach to crafting and labeling these visualization tools will depend on your business and the message you aim to communicate.

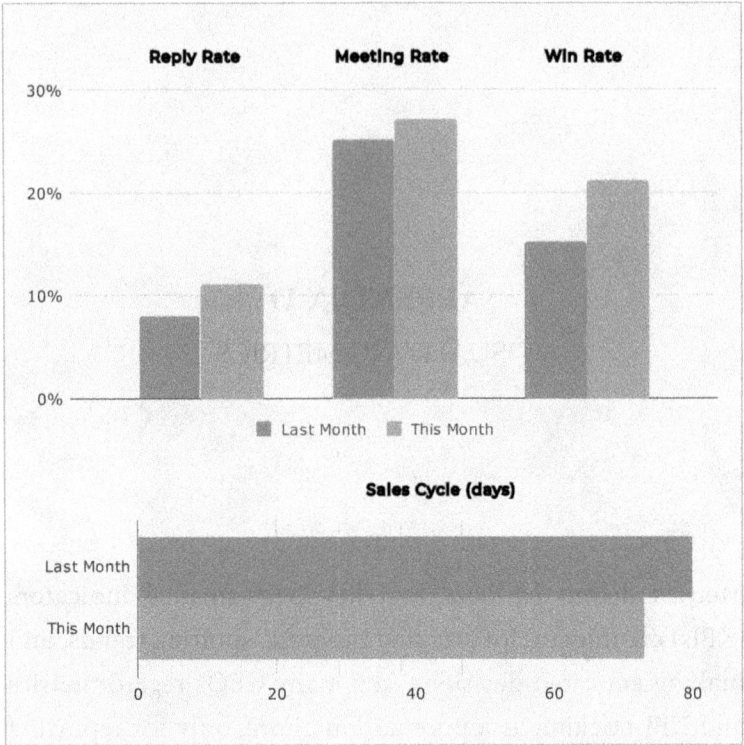

Figure 1: Four Easy Metrics that Deliver Quickly.

Key outbound performance metrics (reply rate, meeting rate, win rate, and sales cycle) provide quick payback when tracked consistently. Even modest improvements across these areas can compound into significant gains in revenue efficiency.

Takeaway: Tracking a small set of simple outbound metrics offers an accessible way to identify progress and accelerate performance.

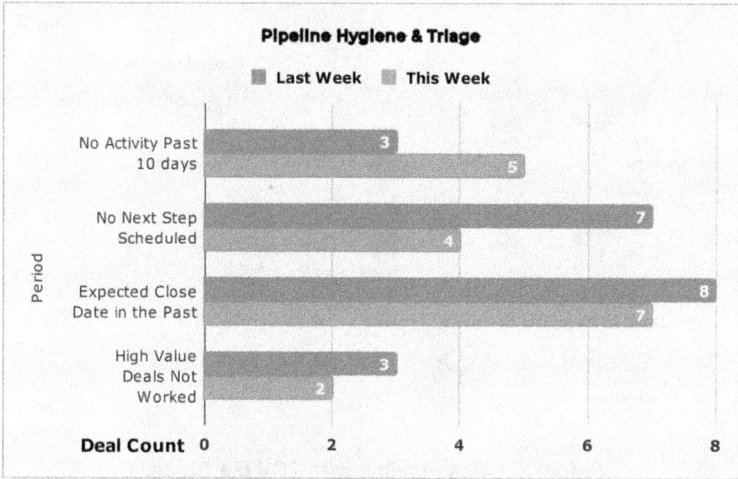

Figure 2: Pipeline Hygiene & Triage Summary.

Regular pipeline reviews are essential to maintaining deal quality and sales discipline. A weekly cadence ensures that every opportunity is actively worked, that stalled or unrealistic deals are cleared out, and that high-value prospects receive appropriate attention. This prevents "zombie" or "fantasy" deals from inflating the pipeline, keeps the team focused on winnable opportunities, and reinforces accountability by linking activity to measurable outcomes. A clean, honest pipeline is the foundation for accurate forecasting and consistent revenue performance.

Takeaway: A disciplined weekly pipeline review turns the pipeline into a tool for focus and truth, rather than wishful thinking.

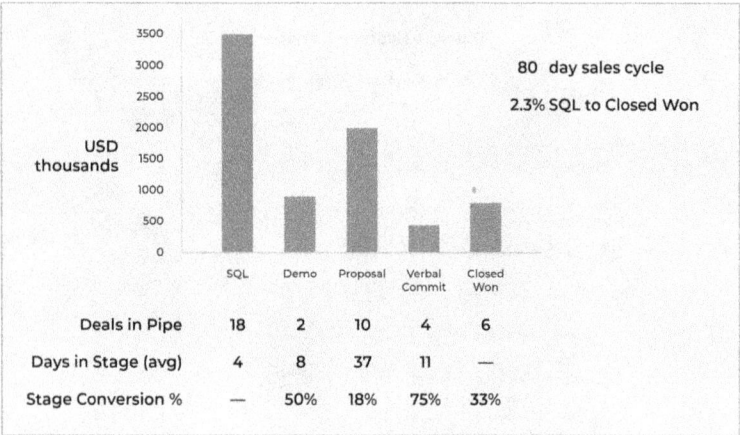

Figure 3: Sales Pipeline with TCV per Stage.

Visualizing your sales pipeline allows you to see where the value is, how long it takes to close deals, and how many deals are in your sales funnel at a glance.

Takeaway: A clear, stage-by-stage pipeline view transforms intuition into evidence. By tracking deal volume, velocity, and conversion, leaders can pinpoint bottlenecks, forecast accurately, and focus coaching where it drives the biggest revenue gains.

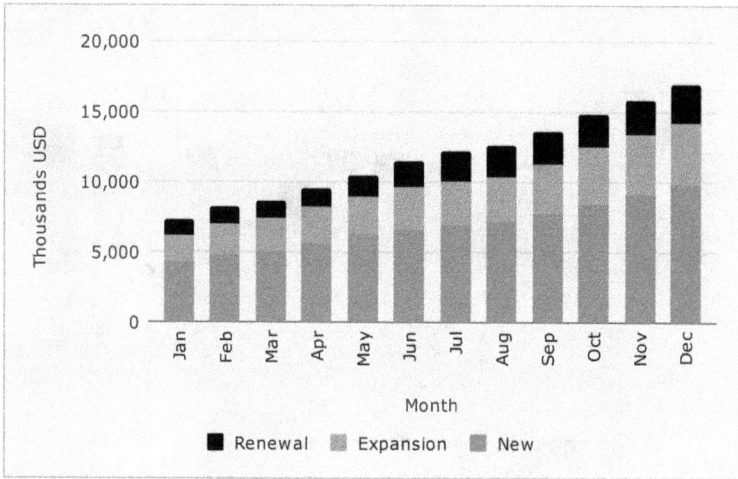

Figure 4: Book of Business.

This figure shows how your ARR (or MRR) Book of Business (BoB) trends. SaaS BoB is a company's recurring revenue from its current customer contracts. This includes recurring new, expansion, and renewal revenue. If a company has a large or growing BoB, it has a robust and (hopefully) stable source of future revenue. A SaaS company's future income and success depend on the health and growth of the BoB. Note that by categorizing the type of revenue, you can easily see how different revenue categories are trending versus the others. For example, you can break down the BoB chart to see revenue trends by product and customer segment.

Takeaway: Tracking your Book of Business shows whether your revenue engine is growing or leaking. A healthy mix of renewal, expansion, and new revenue signals product-market fit, customer satisfaction, and scalable growth. Weakness in any category reveals where retention, upsell, or acquisition needs attention.

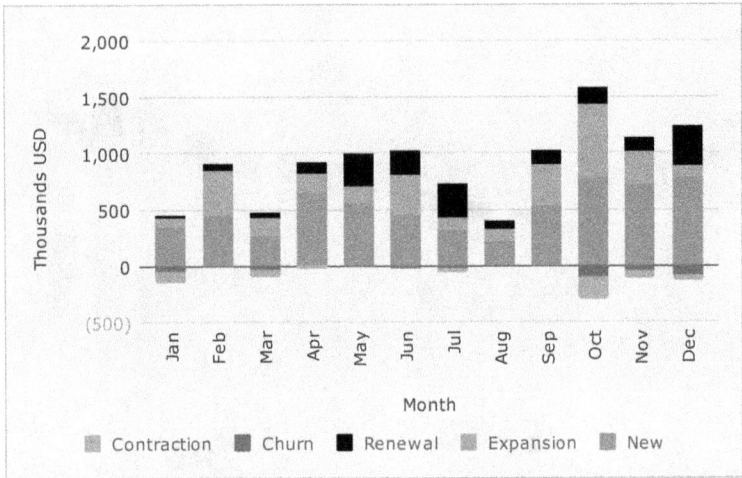

Figure 5: ARR Monthly Bookings.

Unlike the Book of Business, the Bookings chart is not cumulative. This chart shows each month's activity regarding positive and negative changes to your BoB, what category those changes are in (e.g., new business or expansion), and how those categories and the overall change in bookings are trending. This chart helps answer questions such as: Do you see mainly new sales, expansions, or renewals? What were your goals? Do the data match your goals? How do ratios trend over time? Which customers are buying more, and which are not? What is the sales team prioritizing? Are the sales performance or the product areas concerning?

Takeaway: Monthly bookings reveal the true motion behind your growth. By tracking new, expansion, renewal, and churn activity separately, you can see what is driving or dragging ARR. Healthy companies use this view to diagnose sales focus, customer retention, and product-market momentum in real time.

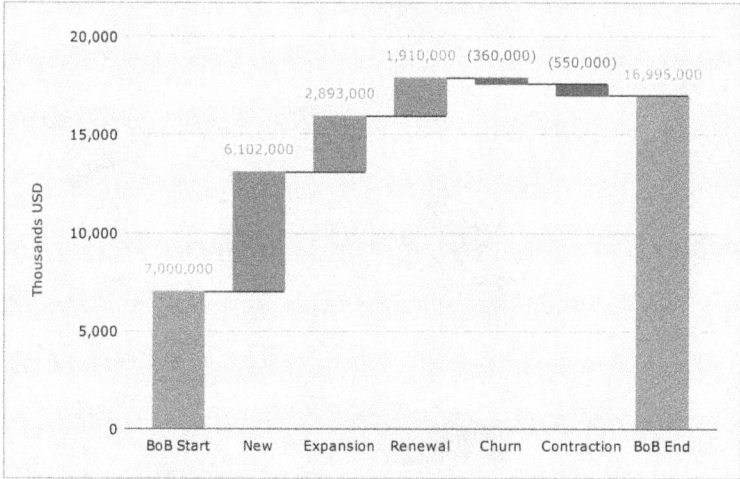

Figure 6: Bookings and BoB Waterfall.

This figure allows you to zoom in on a specific period. The waterfall presentation makes it easier to see what happened over a period of interest. This waterfall is also helpful in understanding the relationship between Bookings and the Book of Business.

Takeaway: A waterfall view turns revenue movement into a story of cause and effect. It shows how new sales, expansions, renewals, churn, and contractions shape total growth. Use it to pinpoint which forces drive or erode your Book of Business over time.

	Jan	Feb	Mar
New Bookings (ACV)	$ 120,000	$ 70,000	$ 23,000
Renewals (ACV)	$ 175,000	$ 340,000	$ 280,000
Lost (ACV)	$ (12,000)	$ (12,000)	—
	Jan	**Feb**	**Mar**
Revenue			
Subscription License Revenue	$ 275,000	$ 270,000	$ 285,000
Usage-based Revenue	$ 150,000	$ 185,000	$ 210,000
Total Subscription Revenue	$ 425,000	$ 455,000	$ 495,000
Professional Services	$ 20,000	$ 25,000	$ 40,000
Other Revenue	$ 12,000	$ 1,000	$ 7,000
Total Revenue	$ 457,000	$ 511,000	$ 542,000
Cost of Goods			
Hosting Expenses	$ 22,000	$ 23,500	$ 26,000
Engineering (Salaries)	$ 14,750	$ 17,000	$ 16,500
Customer Success (Retention)	$ 25,000	$ 22,500	$ 26,500
Direct Costs (Third Party)	$ 2,750	$ 2,750	$ 2,750
Professional Services	$ 37,700	$ 38,500	$ 39,000
Total COGS	$ 101,500	$ 101,500	$ 110,850
Gross Profit	$ 355,500	$ 405,800	$ 431,150
Gross Profit Margin	78%	79%	80%
Operating Expenses			
Marketing	$ 89,000	$ 79,000	$ 90,000
Sales and Customer Success (Sales)	$ 122,000	$ 135,000	$ 140,000
Product Development	$ 125,000	$ 118,000	$ 155,000
General and Administrative	$ 73,000	$ 81,000	$ 83,000
Total Operating Expenses	$ 396,000	$ 413,000	$ 468,000
Operating Profit	$ (40,500)	$ (7,200)	$ (36,850)
Software Amortization	$ (75,000)	$ (74,000)	$ (73,000)
Sales Commissions Amortization	$ (30,000)	$ (35,000)	$ (40,000)
Interest Expense	$ (7,000)	$ (7,000)	$ (7,000)
Net Income	$ (152,500)	$ (123,200)	$ (161,850)

Figure 7: SaaS Income Statement.

This is a standard income statement with bookings and churn layered on top. This is useful for comparing bookings with GAAP (or IFRS) revenue in one chart. Historical financials are essential. Well-founded financials are not only for reporting (though that is essential to running a business). Your financial statements tell the story of your business through numbers. The numbers and your narrative about the company's trajectory must support each other. Your financial performance tells investors and partners what they need to know as your business grows and sells. As the CEO, you also gain insights from financial reports, which show you where to focus on improvements and what's helping or hurting your business.

Takeaway: Your income statement is not just about compliance. It is insight. Use it to connect bookings, churn, and profitability to see how growth translates into real financial health. A disciplined read of your numbers turns data into decisions and strategy into measurable progress.

———

These dashboards are a minimum set for RevOps. Present the data in any form that is easiest for you and your team to consume, and see the trends and implications quickly. Standardize the units and scale of charts that present the same metrics. Never try to make the visual data look better than reality. Don't hurt your business to avoid hurting your ego. It doesn't do your business any good, and experienced board members and investors can see altered-reality charts instantly. Good or bad, the data is there to help you.

APPENDIX E

GLOSSARY OF TERMS

10 × 10 Message Test Loop

A tactical messaging test where you send 10 different message variants to 10 hand-picked targets each (100 total), tracking replies and reactions to identify the most resonant copy before scaling outreach. This is a real-world message and feedback loop, not a focus group.

ACV

Annual Contract Value. Average first-year contract value per deal.

Adoption

Ongoing use of the product's core actions, tracked as leading indicators of retention.

AE

Account Executive. Closing role that owns opportunities to contract.

Anti-ICP

Traits or triggers that predict poor conversion or retention; excluded by default.

ARPA

Average Revenue per Account. Recurring revenue divided by the number of active accounts.

ARPU

Average Revenue per User (or Unit). Total revenue per period divided by the total number of users (or units sold).

ARR

Annual Recurring Revenue. The total value of the annual-normalized recurring revenue book of business.
(see also ARR(e), MRR, RR)

ARR Capacity

Estimated new ARR added annually from the current funnel.
ARR Capacity = SQLm × CR × DS × (12 / SC).

ARR(e)

Effective ARR normalizes non-ARR contracts to 12 months, allowing them to be presented and analyzed together.

ASP

Average Selling Price. Average price per unit or seat, or average deal value divided by units. Normalize the contract length for consistency and comparability.

Attribution

Method to assign credit for outcomes to channels or touches; declares the model used.

Balance Sheet

One of the three standard financial statements, the balance sheet is a snapshot of a company's financial position at a specific time.
(see also Income Statement and Cash Flow Statement)

Book of Business (BoB)

ARR or MRR at a specific date after all changes.
(see Bookings, Churn, and Contraction)

Bookings

Value of contracts signed in a period, regardless of revenue recognition.

Burn Multiple

Net Burn Rate per new dollar of Bookings in a given period. The Burn Multiple measures a company's capital efficiency.

Burn Rate

A company's gross expenditures per period (typically monthly). Sometimes expressed in net terms (i.e., Burn Rate + Revenue).

Buyer Persona

The characteristics of the specific individuals within your ICP who participate in or lead the buying decision.

Buying Committee

Roles that must agree for a deal to close, such as champion, user, economic buyer, IT, and legal.

CAC

Customer Acquisition Cost. Fully loaded sales and marketing costs to acquire one new customer.

CAC Payback

Months to recover CAC from gross margin on new revenue.

Canon

See GTM Canon.

Capacity

Work your team can realistically perform in a period (calls, discovery meetings, demos). You must size campaigns to your capacity, not hope.

Cash Flow Statement

One of the three standard financial statements, the cash flow statement shows the inflow and outflow of cash and cash equivalents from a company's operating, investing, and financing activities over a specific period.
(see also Income Statement and Balance Sheet)

CDP

Customer Data Platform. A CDP pulls data from everywhere your customer interacts, cleans it up, and makes it available so teams can personalize outreach, measure behavior, or automate actions. It's often used alongside or instead of a data warehouse, especially when the company needs real-time, cross-channel customer insights without a heavy data engineering lift.

Champion

An insider who drives the deal, coordinates stake-holders, and wants your solution to win.

Change Log

Central record of definition, field, routing, or metric changes with date, owner, and reason.

Churn Rate

The percentage of customers or revenue lost over a given time period. Measured as either: Customer churn (logos lost) or Revenue churn (dollars lost).

Closed Won

Sales stage or status characterized by a signed or otherwise legally binding contract between a seller and a customer, outlining the provision of defined services in exchange for a defined fee.

COGS

Cost of Goods Sold (sometimes called COS or Cost of Sales, depending on your business model). This term has a specific definition in a company's accounting.

Cohort

A group that started in the same period (e.g., leads first seen in June). Used for apples-to-apples performance reads. Often used in marketing and also valuable for sales pipeline analysis, especially for identifying ICPs.

Cohort Retention

Percentage of a starting cohort's ARR that remains after N periods.

Contraction

Decrease in ARR from existing customers during a period.

Conversion Chain

Meeting Rate × Opp Rate × Win Rate.
Equivalent to weighted full-funnel conversion efficiency.

Conversion Rate

The percentage that advances from one stage to the next in a period.

Coverage

Pipeline value divided by the sales target for a period (e.g., 3× coverage for the quarter). Coverage without high-confidence stages is a vanity metric. The best practice is to use all qualified pipeline scheduled to close in that period, not the entire open pipeline (e.g., if the Q4 target is $1M and you have $3M in opportunities with close dates in Q4 → coverage = 3×).

CRM

Customer relationship management software (e.g., HubSpot, Salesforce).

CS/CSM

Customer Success or Customer Success Manager. Post-sales role that drives onboarding, adoption, renewal, and expansion. Some sources may refer to CX (customer experience).

Customer Journey

The customer journey is how the buyer advances toward a decision. See also Sales Process.

Data Dictionary

The controlled list of metric and field definitions used across tools and reports.

Data Warehouse

A central database that consolidates data from multiple systems for analysis and business intelligence.

Disqualify

Remove a lead or opportunity permanently for a defined reason code.

Dollar Churn

ARR lost in a period from existing customers.

Economic Buyer

The person who controls the budget and signs the agreement.

ETL / ELT / Reverse ETL

ETL (Extract, Transform, Load)
Data is **extracted** from source systems (like CRMs or apps), **transformed** into a clean, consistent format, and then **loaded** into a data warehouse.
→ *Used when data needs heavy cleaning or modeling before storage.*

ELT (Extract, Load, Transform)
Data is **extracted** and **loaded first** into the warehouse, then **transformed** there.
→ *Used in modern cloud warehouses (like Snowflake or BigQuery) where transformation is faster and cheaper at scale.*

Reverse ETL
Takes **modeled data from the warehouse** and **syncs it back** into operational systems (like HubSpot, Salesforce, or ad platforms).
→ *Used to activate insights — for example, sending product-usage scores or churn risk back to sales or marketing tools.*

Exit Criteria

Required, observable conditions to move a deal to the next stage; recorded as pass or fail.

Expansion Rate

Expansion ARR in period divided by starting ARR.

Forecast Accuracy

(Actual – Forecast) / Actual.

Forecast Bias

Directional error tendency over time (optimistic or conservative).

Forecast Categories

Standard pipeline buckets such as Commit, Best Case, Pipeline, each with entry criteria.

GAAP

Generally Accepted Accounting Principles. The accounting rules and standards used in the United States for preparing financial statements. GAAP emphasizes consistency, comparability, and detailed disclosure. It governs how revenue is recognized, how expenses are matched, and how assets/liabilities are valued. See also IFRS.

Give-Get

Concession rule pairing any discount or favor with a reciprocal customer commitment.

Gross Margin

(Revenue – Cost of Goods Sold) / Revenue.

GRR

Gross Revenue Retention measures how much recurring revenue you keep from existing customers after accounting for churn and downgrades, but ignores expansions.
Formula: (Starting ARR – Contractions – Churn) ÷ (Starting ARR). This calculation excludes expansions.
(See also NRR)

GTM

Go-To-Market. A coordinated system of processes, tactics, and insights that brings a product to market and generates revenue. GTM is not a department; it's a cross-functional revenue engine connected by RevOps.

GTM Canon

The baseline set of GTM metrics, practices, and stage gates such as ICP clarity, message testing, pipeline hygiene, handoff SLAs, and forecast discipline. The "canon" can vary slightly or broadly from company to company.

GTM Fit

Demonstrated evidence that messaging, offers, and channels consistently generate qualified conversations and close deals without the founder doing everything manually. It is the precondition for scaling.

GTM Maturity Stage

GTM maturity is your company's position on the GTM maturity ladder. This is about the business, not a deal pipeline stage. Below is how we think about a company's different growth stages.

- Zero to Something
- Selling Before Scaling
- Hiring Without Breaking
- Scaling With Confidence

Handoff

The moment ownership changes. For example, MQL → SDR, SDR → AE, AE → CSM. Record handoffs, which require a next step on the receiving side.

Hit Rate

Meetings that become qualified opportunities, divided by total meetings.

ICP

Ideal Customer Profile. Segment that buys fast at acceptable economics and renews; defined by traits and triggers.
ICP is typically the account (company), while the person you want to talk to is described by the Persona.

ICP Fit Score

Scored measure of how well a lead matches ICP traits and triggers.

Identity Key

Unique field used to match entities across systems (e.g., account_id, email).

Identity Resolution

Matching people or accounts across systems using keys to form a single record.

IFRS

International Financial Reporting Standards are the accounting framework used by most countries outside the United States. IFRS are principles-based, meaning they provide broad guidelines rather than highly prescriptive rules. Designed for global consistency so investors can compare companies across borders.
(See also GAAP)

Inbound Marketing

The process of attracting prospects through content, SEO, ads, referrals, or social proof. Requires validated messaging, clarity on ICP, and the ability to qualify interest quickly.

Income Statement

One of the three standard financial statements, the income statement, also known as the profit and loss statement (P&L), shows a company's revenues, expenses, and net income over a specific period (see also Balance Sheet and Cash Flow Statement).

Infosec

Information security review of products and processes during procurement.

iPaaS

Integration Platform as a Service. Tool for automating data syncs and workflows across apps.

Lead

Person or account that has shown interest but is not yet qualified against ICP and intent.

Lead Leak

The preventable loss of a qualified hand-off between funnel steps due to process, routing, or tooling gaps, not buyer rejection.

Lead Routing

Rules that assign new records to owners based on segment, territory, or round-robin.

Lead Scoring

A model that ranks lead priority using firmographic and behavioral signals.

Legal Review

Contract term review and redlines managed by legal or procurement.

Logo Churn

Count of customers lost in a period, regardless of dollar value.

LTV

Life Time Value (sometimes CLV; Customer Lifetime Value). Present value of gross margin over expected life; simple model: (ARPA × Gross Margin %) / Monthly Churn.
Sometimes also called CLTV.

MAP

A dated checklist shared with the customer that defines the steps from proof-of-value to signature to go-live. A mutual action plan (MAP) signals intent to buy and an agreed-upon roadmap regarding what is required to close the deal.

Marginal Cost

The additional COGS (or COS) incurred by adding each new user or unit. In other words, it is the cost of producing one additional unit (product or service) of output.

Meeting Rate

Meetings booked divided by targeted accounts or outbound touches.

Monthly Operating Review

Deeper checkpoint on conversion, pricing, hiring, and channel bets; closes the loop on experiments.

Motion

A specific go-to-market strategy such as Outbound, Inbound, or PLG.

MQL

Marketing Qualified Lead. Lead that meets marketing's engagement and fit thresholds and is ready for sales review.

MQL→SQL Rate

SQLs divided by MQLs for the period.

MRR

Monthly Recurring Revenue. Monthly contracted recurring revenue.

MSA

Master Service Agreement. A contract that defines general terms between parties.

Net New ARR

New ARR won in a period minus churn and contraction for that period.

No-Show Rate

Percentage of booked meetings that do not occur.

NRR

Net Revenue Retention measures how much recurring revenue you retain from existing customers after accounting for churn, downgrades, and expansions (upsells/cross-sells).
Formula: Starting ARR + Expansions – Contractions – Churn ÷ Starting ARR
This calculation includes expansions.
(See also GRR)

Opportunity

Qualified selling motion with value, stage, owner, and next step.

Opportunity Yield

Opportunities ÷ Meetings.

Outbound Sales

A GTM tactic where a company initiates contact with prospects (via email, phone, LinkedIn, etc.) to generate a pipeline. Often validates messaging and ICP faster than inbound.

Painted Door Test (aka, Fake Door)

A demand test where you present a real CTA to a not-yet-built feature or offer, capture clicks and hand-raises, and decide whether to build based on measured interest.

Pilot/POC

Proof of Concept. Timeboxed test to de-risk a purchase by proving value in the customer environment. POCs frequently stall or fail to convert to a full contract due to poorly defined goals and expected outcomes.

Pipeline

The total value of active deals that could potentially close. Often segmented by stage and weighted by probability.

Pipeline Coverage

Open pipeline for the target period divided by quota or ARR target. A typical health benchmark: 3–5x quota.

Pipeline Stage

A step in your sales process with clear, verifiable exit criteria. Keep pipeline stages to the minimum needed for pipeline hygiene and analysis. Label the stages with clarity. Ensure you define clear exit criteria for each stage. Some operators prefer to build pipeline stages around internal activities, while others prefer customer-centric activities. Choose the best method for your business needs.

Pipeline Stage Age

The number of days a deal has been in the current stage. Monitor stage aging to spot stuck deals and enforce management reviews. If a specific pipeline stage has a long average age, consider splitting it into two or more stages to find the bottleneck.

Pipeline Stage Exit Criteria

The non-negotiables to move forward (e.g., economic

buyer confirmed and next step scheduled). Confirm the exit criteria or do not advance the deal.

Pipeline Velocity

(Number of Opps × Win Rate × ACV) / Sales Cycle (months).

Play

A repeatable tactic that defines exactly who you're targeting (ICP), why now (trigger), what you're saying (message), and how you're delivering it (channel).

PLG

Product-Led Growth. A go-to-market approach where product usage drives acquisition, activation, and expansion. Often includes freemium models or free trials. Requires strong UX and onboarding without sales intervention.

PMF

Product-Market Fit. The state in which a product solves a meaningful problem for a specific market segment in a way that drives strong usage, retention, referrals, and revenue.

PMM

Product Marketing Manager (or Product Marketing).
Role: They own positioning, messaging, personas,
competitive intel, and enablement.
In SaaS GTM: PMM bridges product → sales/mar-
keting, ensuring ICP pain, value propositions, and
assets are crisp.

Positioning

How you frame your product in the market relative to
alternatives.

Price Realization

How much of your list price you actually capture at
close and over the contract, after discounts, conces-
sions, credits, and non-standard terms.

Procurement

The buyer's purchasing function that governs vendor
risk, terms, and approvals.

Proof of Value (evidence)

Concrete, verifiable artifacts that reduce buyer risk

and support the value claim. Examples include a named case study, pilot outcome, ROI model, benchmark, security attestation, or customer quote tied to a metric. Proof of value should be specific, attributable, recent, and mapped to the buyer's key objections.

QBR

Quarterly Business Review. Strategic review of performance vs plan, resourcing, and roadmap; resets stage gates.

Quota Attainment

Actual bookings divided by quota for the period.

RACI

Matrix clarifying who is **Responsible, Accountable, Consulted, and Informed** for a decision or process.

Recycle

Move a lead or opportunity out of the active queue with a scheduled next review.

RevOps

Revenue Operations. Aligns GTM teams with systems, processes, and data. RevOps ensures that metrics are visible and actionable, and that systems support execution.

RevOps Backbone

Minimum set of definitions, fields, SLAs, capture, and scorecards that instrument the funnel end-to-end; stable core, modular edges.

RR

Recurring Revenue. A generic term that could mean MRR, ARR, or other recurring revenue metric.

Runway

A measure of how many months a company can operate with the current cash on hand.
Runway = Cash / Burn Rate

SAL

Sales Accepted Lead. MQL reviewed and accepted by sales within the SLA for follow-up.

Sales Cycle

Days from the first qualified meeting to Closed Won. It is helpful to track deals overall, by segment, and by ranges (revenue bands) of annual contract value (ACV). Separately track the time to the first qualified meeting, a marketing metric. RevOps also supports your marketing activities.

Sales Cycle

Median days from SQL to close-won. Many operators use the mean days.

Sales Process

The sales process is how your team advances the deal.
(See also Customer Journey)

Sales Velocity

How quickly the pipeline turns into revenue.
(Number of Deals) × (Average Deal Size) × (Win Rate) ÷ (Sales Cycle).

Sandbox

Non-production environment for testing changes before rollout.

Scorecard

Tactical reports showing progress against weekly or monthly targets.

SDR

Sales Development Representative. Prospecting role that books qualified meetings; sometimes called a Business Development Representative (BDR).

SEP

Sales Engagement Platform. A tool for sequencing touches, email, and call logging.

Sequence

Timed set of touches across channels to engage a prospect.

SLA

Service Level Agreement. A formal or informal
agreement between two parties to take specific
actions within a specific period.
The SLA is a time-bound rule for handoffs (e.g., from
marketing qualified lead (MQL) → sales development representative (SDR) within 24 hours, SDR →
account executive (AE) within four hours). Ensure
that you trigger alerts for missed SLAs.

SLA Attainment

Percentage of items responded to within the SLA
window.

SOW

Statement (or Scope) of Work. A contract that defines
scope, deliverables, and timelines for services.

Sprint

A fixed two-week period to test measurable go-to-
market (GTM) outcomes. These are experiments you
run that challenge your assumptions. Plan one sprint
at a time. Don't confuse GTM sprints with software
sprints.

SQL

Sales Qualified Lead. Lead that meets the agreed discovery criteria and enters the opportunity pipeline.

SQLm

Sales-qualified leads per month.

SSOT

Single Source of Truth. The system or model that is authoritative for a specific entity (Account, Contact, Opportunity, Product).

Stage Age (pipeline)

Number of days a given opportunity has remained in its current stage.

Stage Aging (pipeline)

Median days opportunities spend in a given stage.

Stage Gate

Predefined test of a new message, channel, hire, or tool. Stage gates serve as readiness checkpoints that must succeed before adding resources, volume, or budget. This drives evidence over opinions. Seriously reconsider a decision to move forward when the data doesn't support it.

Stage, GTM Maturity

Organizational growth phase (e.g., Zero to Something, Selling Before Scaling, Hiring Without Breaking, Scaling With Confidence).

TCV

Total Contract Value. The total recurring and non-recurring revenue expected from a customer contract, including multi-year obligations.

TFR

Time to First Response. Time from inbound signal to first human response.

Trigger

Recent, observable event that increases buy probability (new funding, new exec, tool churn).

TTFV

Time to First Value. Time from contract to the customer achieving an agreed first outcome.

UTM Parameters

Standard URL tags that record campaign source, medium, and content.

Value Proposition

Short statement of the outcome you deliver for a specific buyer.

Voice of Customer

Structured program to capture insights from customers and prospects.

Weekly Revenue Review

Cross-functional meeting that inspects pipeline, leaks, SLAs, and actions using two scorecards.

Win Rate

Closed won divided by qualified opportunities, defined consistently across the organization.

APPENDIX F
WORKSHEETS

You can download templates, worksheets, checklists, GTM-focused AI bot assistants, and planning tools referenced in the book to help you apply the content directly to your go-to-market. The materials are updated periodically. Register for updates on our website's Resources page.

Access Companion Resources

mirmeridian.com/resources

How to Use The Worksheets

This is your workbench. Each worksheet creates a decision or artifact your team will use daily. Execute one worksheet for each meeting. Decide, document, and move one metric per two-week sprint.

Complete the **Quick Start Pack** in order WS-01 to WS-05. Then, based on your team's gaps, select one or two Topic Pack worksheets every two weeks.

Review the results with your team's weekly or bi-weekly. Use pass or fail and keep or kill rules to decide on the next steps.

Note where you recorded this worksheet data so your team can find it. Many worksheets create an output that other worksheets may need.

The worksheets follow a standard format for consistency. Not all fields and rubrics always apply. Skip sections if they are irrelevant to your current stage, process, or test.

Each worksheet is one decision, one artifact, and one measurable outcome.

On the next pages, you'll find a quick-reference index of all worksheets—each with its purpose and use case. →

———

Worksheet Index

QUICK START

WS-01 GTM SNAPSHOT

Supports: Ch1, Ch2 | Timebox: 60 min | Participants: CEO, Sales, Marketing, CS, RevOps

Purpose: Align the team on what is true now across ICP, message, channels, process, proof, pricing, and capacity. Create a single-page snapshot that anchors the next two-week sprint and clarifies where to focus effort.

Use When: Kicking off a new GTM push or entering a new stage gate in your growth cycle.

WS-02 GTM MATURITY STAGE ASSESSMENT

Supports: Ch2, Ch17 | Timebox: 60 min | Participants: CEO/Founder, Executive Team, RevOps

Purpose: Score your current GTM Maturity Stage, define missing evidence, and select one weekly Bet per team to improve red or yellow areas. Produce a clear stage scorecard and change log.

Use When: After WS-01 and before large resource or budget decisions; repeat monthly.

WS-03 ICP GRID AND ANTI-ICP: TRAITS, TRIGGERS & EVIDENCE

Supports: Ch4 | Timebox: 60 min | Participants: Sales, Marketing, CS, RevOps

Purpose: Define the buyer profiles most likely to close quickly at full price and capture their behavioral triggers with proof from real deals. Identify anti-ICP red flags to avoid wasted outreach.

Use When: After WS-02 or anytime you see inconsistent win rates or discounts across segments.

WS-04 MESSAGE TEST GRID: VARIANTS AND DECISIONS

Supports: Ch5 | Timebox: 60 min | Participants: Product, SDR Lead, AE Lead, RevOps

Purpose: Test message variants using buyer language and numeric thresholds instead of opinions. Build a repeatable A-B-C grid that turns feedback into data.

Use When: After WS-03 and anytime reply rates drop or new segments emerge.

WS-05 FIRST SALES PROCESS & MAP

Supports: Ch8, Ch10 | Timebox: 90 min | Participants: Sales, RevOps, Top AEs, HR

Purpose: Document the whole path from first touch to close with clear exit evidence, SLAs, and a buyer-facing Mutual Action Plan (MAP). Create one teachable, auditable process.

Use When: Before hiring additional reps or formalizing forecasting; refresh every two weeks during early scale.

MESSAGING

WS-06 PROOF OF VALUE LIBRARY

Supports: Ch5, Ch10 | Timebox: 45 min | Participants: Product, Sales, CS

Purpose: Catalog every proof-of-value asset—case studies, demos, testimonials—and map each to its relevant pipeline stage and owner. Maintain a living inventory that drives confidence and conversions.

Use When: Before refresh cycles or enterprise pushes; revisit quarterly.

WS-07 OBJECTION LIBRARY

Supports: Ch5, Ch10 | Timebox: 60 min | Participants: Product, Sales

Purpose: Turn common objections into structured responses supported by proof and a clear next step.

Build a searchable objection library for training and outbound sequencing.

Use When: Before onboarding new reps or after patterns of stalled deals appear in win/loss reviews.

OUTBOUND

WS-08 TWO-WEEK OUTBOUND SPRINT PLAN

Supports: Ch6 | Timebox: 45 min | Participants: SDR Lead, AE Lead, RevOps

Purpose: Run a focused two-week outbound sprint with defined activity targets, message, and review cadence. Produce a short plan that measures what worked and what didn't.

Use When: At the start of every outbound sprint or when outbound consistency declines.

WS-09 LEAD LIST QUALITY CHECKLIST

Supports: Ch6, Ch4 | Timebox: 30 min | Participants: SDR Lead, RevOps

Purpose: Enforce standards for list accuracy and ICP fit before campaigns go live. Identify and correct weak data or missing triggers before wasting send volume.

Use When: Before every outbound launch or vendor import.

WS-10 OPENER VARIANTS TRACKER

Supports: Ch6, Ch5 | Timebox: 30 min | Participants: SDR Lead, Product, RevOps

Purpose: Track performance of opener lines or hooks across sprints to decide which to keep or kill. Maintain a simple log connecting message changes to results.

Use When: During each outbound sprint review or when reply rates fluctuate.

INBOUND AND PLG

WS-11 FUNNEL ENTRY POINTS AND SLAS: ROUTES AND RESPONSE TIMES

Supports: Ch7, Ch9 | Timebox: 60 min | Participants: Marketing, Sales, CS, SDR Lead

Purpose: Define every funnel entry point and enforce clear routing, response-time SLAs, and ownership. Reduce leakage caused by slow or unclear handoffs and ensure every lead receives the right action at the right speed.

Use When: Before launching new campaigns, adding

inbound channels, or anytime SLA misses or lead leakage appear in reviews.

WS-12 ACTIVATION TRIGGERS MAP: PREDICTING PQLS

Supports: Ch7 | Timebox: 60 min | Participants: Product, Data, Sales, RevOps

Purpose: Identify product usage events that reliably predict buying intent. Define thresholds, signal strength, and the required next touch so Product Qualified Leads (PQLs) are detected and acted on consistently.

Use When: Once you have a working product with usage data and want to connect activation behavior to sales or assist motions; revisit monthly in PLG models.

WS-13 SALES-ASSIST HANDOFF CARD

Supports: Ch7, Ch10 | Timebox: 45 min | Participants: Sales, Product, CS, RevOps

Purpose: Standardize when and how Sales assists product-led users. Define eligibility criteria, the assist talk track, and the next concrete step so help is timely, relevant, and measurable.

Use When: You run a free trial or freemium motion

at scale and need to decide which users receive human outreach, how it happens, and what success looks like.

PROCESS & REVOPS

WS-14 REVOPS FIELD DEFINITIONS

Supports: Ch9, Ch18 | Timebox: 60 min | Participants: RevOps, Sales, Marketing, CS

Purpose: Create a shared understanding of every CRM and reporting field—its purpose, owner, and refresh schedule. Eliminate duplicate and unclear fields to protect data trust.

Use When: Before any process change, new tool implementation, or quarterly data hygiene audit.

WS-15 STAGE EXIT CRITERIA: EVIDENCE TO ADVANCE

Supports: Ch10, Ch18 | Timebox: 60 min | Participants: CEO/Founder, Sales, RevOps

Purpose: Define the concrete evidence required for each pipeline stage exit. Ensure stage accuracy, faster forecasting, and consistent buyer progression.

Use When: Before hiring, revising forecast models, or redefining sales stages.

WS-16 ROUTING SLAS AND RECYCLE RULES

Supports: Ch9, Ch7 | Timebox: 60 min | Participants: RevOps, SDR Lead, Marketing

Purpose: Set routing speed targets, disqualify logic, and recycle rules to keep leads fresh. Improve follow-up speed and accountability across teams.

Use When: After WS-11 or ahead of campaign launches where inbound and outbound intersect.

PRICING

WS-17 PRICING GIVE-GET BANDS & DISCOUNT GUARDRAILS

Supports: Ch11 | Timebox: 60 min | Participants: CEO or CRO, Sales, Finance, RevOps

Purpose: Establish discount guardrails with defined give-get rules to protect margins. Set approval thresholds and ensure trade-offs are consistent.

Use When: At the start of each quarter or before revising pricing or discount policies.

WS-18 DISCOUNT APPROVAL FLOW

Supports: Ch11, Ch18 | Timebox: 45 min | Participants: Sales, Finance, RevOps

Purpose: Define a clear flow for pricing exceptions and approvals. Speed up approvals while reducing uncontrolled discounts.

Use When: Alongside WS-17 or before adding new discount tiers or policies.

WS-19 PACKAGING TEST CARD

Supports: Ch11, Ch5 | Timebox: 60 min | Participants: Product, Sales

Purpose: Test new packaging or tiering hypotheses with clear metrics and thresholds: record offer details, audience, test window, and outcomes.

Use When: Before changing pricing, bundling, or packaging strategy.

HIRING

WS-20 AE SCORECARD: SKILLS, BEHAVIORS, EVIDENCE

Supports: Ch12 | Timebox: 60 min | Participants: CEO/Founder or CRO, Sales, HR

Purpose: Hire AEs using evidence-based criteria—skills, behaviors, and performance proof. Build consistency across hiring decisions.

Use When: Before interviews or when updating AE hiring standards.

WS-21 SDR SCORECARD: ACTIVITY, QUALITY, LEARNING VELOCITY

Supports: Ch12 | Timebox: 60 min | Participants: SDR Lead, HR

Purpose: Evaluate SDR candidates on the quality of activity, adaptability, and learning speed using structured tasks. Build a repeatable hiring process that predicts ramp success.

Use When: Before interviewing new SDRs or training a new cohort.

WS-22 "30-60-90" DAY PLAN: ENABLEMENT AND GO / NO-GO

Supports: Ch12, Ch10 | Timebox: 45 min | Participants: Manager, New Hire

Purpose: Design a structured 30-60-90 ramp plan with clear milestones and go/no-go reviews. Set expectations and track progress.

Use When: On a new hire's first day or during onboarding redesign.

WS-23 RAMP MATH AND CAPACITY: QUOTA, PIPELINE NEEDS, PAYBACK

Supports: Ch12, Ch3 | Timebox: 60 min | Participants: Sales, Finance, RevOps

Purpose: Model hiring needs and ramp timelines using real funnel math. Quantify how many reps, leads, and deals are required to hit revenue targets.

Use When: During quarterly planning or before adding new headcount.

VOICE OF CUSTOMER

WS-24 VOC CAPTURE CARD: CONTEXT, QUOTES, PAIN, PROOF

Supports: Ch12, Ch5 | Timebox: 30 min per interview | Participants: any interviewer

Purpose: Standardize how your team captures voice-of-customer insights—context, verbatim quotes, pain points, and proof of value—so interviews become comparable data instead of anecdotes.

Use When: During ongoing customer interviews or feedback cycles; review monthly.

WS-25 TAXONOMY AND THEMES: CODE AND SPOT PATTERNS

Supports: Ch13 | Timebox: 60 min | Participants: Product, Sales, CS

Purpose: Build a shared vocabulary for customer feedback by tagging, coding, and ranking themes from capture cards. Expose the patterns driving churn, adoption, or delight.

Use When: Monthly or after at least 15 interviews.

WS-26 PRIORITIZATION SCORING MATRIX

Supports: Ch13, Ch15 | Timebox: 60 min | Participants: ELT, Product, GTM Leads

Purpose: Score competing opportunities using weighted criteria to surface the top 5 bets most likely to create impact.

Use When: During quarterly planning or roadmap reviews.

INVESTORS, BOARD, AND ADVISORS

WS-27 DECISION BRIEF: CONTEXT, OPTIONS, RISKS, RECOMMENDATIONS

Supports: Ch14, Ch17 | Timebox: 45 min | Participants: Owner and Reviewers

Purpose: Frame a high-impact decision on one page —summarizing context, options, risks, and recommendations—to speed up executive or board alignment.

Use When: Any time a major strategic choice requires clear framing or stakeholder input.

WS-28 VRMON RISK CARD

Supports: Ch14, Ch15 | Timebox: 45 min | Participants: ELT

Purpose: Track material risks and mitigation actions using the VRMON structure (Visible Risks, Mitigations, Options, Next step). Maintain a live risk register.

Use When: During monthly ELT or board risk reviews.

WS-29 INVESTOR UPDATE TEMPLATE

Supports: Ch14, Ch18 | Timebox: 60 min | Participants: CEO/Founder, Finance, RevOps

Purpose: Deliver honest, concise investor updates that combine facts, learnings, and clear asks for help.

Use When: Monthly or quarterly before board meetings.

SCALING

WS-30 GROWTH THESIS ONE-PAGER: WHY SCALE NOW?

Supports: Ch15, Ch17 | Timebox: 60 min | Participants: ELT

Purpose: State the evidence-based case for scaling a motion. Document why, what risks exist, and what thresholds must be met before expansion.

Use When: After repeatable wins or before committing new budget.

WS-31 CAPACITY MODEL: SOM AND SQL MODEL

Supports: Ch3, Ch17 | Timebox: 60 min | Participants: RevOps, Finance, Sales

Purpose: Calculate bottom-up ARR capacity using real funnel inputs (Reach, Meeting Rate, Pipeline Conversion, Win Rate, Deal Size). Create a defensible forecast tied to execution limits.

Use When: During quarterly planning or before scaling sales headcount.

WS-32 CHANNEL PORTFOLIO GUARDRAILS: CAPS, CUT RULES, MIX

Supports: Ch15 | Timebox: 45 min | Participants: ELT, Marketing, Sales

Purpose: Define channel caps, cut rules, and payback targets to prevent sprawl and wasted spend.

Use When: Quarterly or whenever budget allocation shifts.

REPORTING

WS-33 CEO DASHBOARD SETUP: 7 TO 10 KPIS

Supports: Ch18, Ch17 | Timebox: 60 min | Participants: CEO, RevOps, Finance

Purpose: Build a one-page, weekly-refreshed dashboard that drives executive decisions instead of vanity reporting.

Use When: During operating-cadence setup or dashboard redesign.

WS-34 METRIC CANON: DEFINITION, FORMULA, SOURCE, OWNER, REFRESH

Supports: Ch18 | Timebox: 60 min | Participants: RevOps, Finance

Purpose: Document every core metric's formula, source, owner, and refresh cadence to create one source of truth.

Use When: After WS-33 or whenever metric definitions diverge across teams.

WS-35 FORECAST RULES AND COMMIT CRITERIA: LANES AND ACCURACY CHECKS

Supports: Ch18, Ch10 | Timebox: 60 min | Participants: Sales, Finance, RevOps

Purpose: Define forecast lanes and commit rules tied to validated buyer evidence. Improve forecast accuracy and accountability.

Use When: Monthly or before quarterly forecast cycles.

ARTIFICIAL INTELLIGENCE

WS-36 AI POLICY ONE-PAGER: TRUTH, PRIVACY, HUMAN-IN-THE-LOOP

Supports: Ch19 | Timebox: 60 min | Participants: CEO, Legal, CISO, RevOps

Purpose: Set the foundational principles for responsible AI use—truth, privacy, and human oversight. Define acceptable use cases, limits, and approval process.

Use When: Before any AI tool deployment or data integration.

WS-37 PROMPT LIBRARY: GOAL, INPUTS, CONSTRAINTS, EXAMPLE

Supports: Ch19, Ch5, Ch6, Ch7, Ch9 | Timebox: 30 min per prompt | Participants: Function Lead, RevOps, CISO

Purpose: Create and test short, reliable prompts per function with a clear structure: goal, inputs, constraints, and example. Build a vetted prompt library.

Use When: After WS-36 or when standardizing AI-assisted tasks across teams.

WS-38 EVALUATION HARNESS: SAMPLE, SCORECARD, RED TEAM

Supports: Ch19 | Timebox: 60 min | Participants: Function Lead, RevOps, CISO

Purpose: Test AI outputs with controlled samples and scorecards to measure accuracy, reliability, and bias. Provide a repeatable evaluation framework.

Use When: Before AI rollout and quarterly for model validation.

WS-39 ANOMALY ALERTS PLAYBOOK: SLA MISS, STAGE CREEP, DISCOUNT DRIFT

Supports: Ch18, Ch19 | Timebox: 60 min | Participants: RevOps, Function Leads

Purpose: Define data and process alerts with thresholds and owners to detect issues early (SLA misses, forecast drift, discount anomalies).

Use When: After WS-33 and WS-34 or when building automated RevOps monitoring.

ACKNOWLEDGMENTS

We wrote this book while running a business. That wouldn't have been possible without each of us, John, Ash, and Carlos, showing up, carrying the load, writing and editing late into the night. We acknowledge and appreciate the time, thought, and effort each of us contributed to get this book across the finish line.

Muffie Humphrey generously and precisely shared her expertise in self-publishing. Her guidance helped us avoid needless friction and publish faster, smarter, and better.

Benjamin Tu helped shape the structure and flow with his early feedback as a reader and collaborator.

We thank Stew Friedman, Cory Bray, Sandy Sharp, Mark DeSantis, Robin Hanson, Nick Costides, and Tony Scelzo for their thoughtful reviews. Your input strengthened the final product.

To our families: Thank you for your patience during the long hours and missed dinners. You held the fort so we could hold the pen.

To everyone who encouraged us along the way (friends, clients, partners, and peers), thank you for sharing your thoughts, questions, and energy. You helped us sharpen the message.

Finally, a special thank you to the entire Mir Meridian team. Your support made this possible.

We wrote this book for founders, but we didn't write it alone.

ABOUT THE AUTHORS

John Braze, Ash Archibald, and Carlos Vadillo are the founding team behind Mir Meridian. They've sold, built, rebuilt, grown, advised, and rescued more than their fair share of start-ups. They didn't write this book from the cheap seats. John, Ash, and Carlos are hands-on growth partners for start-ups, growth-stage, and midmarket businesses building go-to-market systems that work.

The authors wrote this book after conducting hundreds of CEO strategy sessions, sending cold emails, managing messy CRMs, and working late nights to untangle chaos. They've been in the weeds and get things done. That's where GTM clarity happens.

John Braze is a former submariner who brings deep operational discipline to chaotic growth. He gets to the point, telling founders the hard truths they need to know, particularly when goals aren't aligned, sales strategies are misleading, or messaging is off. When he says, "The facts don't lie," you'll believe him.

Ash Archibald is an analytical engine. Equal parts calm and conviction, Ash helps companies see what's happening inside their GTM and what's not. His background in data, operations, and start-up execution makes him a rare breed. If dashboards and data visualization give you anxiety, working through the methods developed by Ash will be your antidote.

Carlos Vadillo is the voice of the customer. Carlos's background in sales, marketing, and design helps him understand customers, giving him a unique perspective. He has a gift for translating complexity into clarity and turning "founder speak" into buyer action. His advice is street-tested. Carlos formerly worked as an investment banker and cofounded an ag-tech start-up.

Together, they run Mir Meridian, not as advisors on the sidelines, but as embedded partners who help companies build the system before the storm. They've supported technical founders navigating their first sales call, Series A teams facing their first real forecast, and pre-IPO operators trying to scale revenue without burning out the team.

This book isn't their side hustle. It's the field guide they wish they and every founder had when things got real. Eventually, every founder faces a time when effort alone isn't enough, and what happens next depends entirely on the system you've built.

You're not alone if you've ever thought, "Why is this so hard?" The authors are already in your corner. This book makes it official.

linkedin.com/company/mirmeridian
x.com/mirmeridian